EXPLODING THE GENE MYTH

EXPLODING THE GENE MYTH

How Genetic Information Is Produced and Manipulated
by Scientists, Physicians, Employers,
Insurance Companies, Educators, and Law Enforcers

Ruth Hubbard and Elijah Wald

Beacon Press · Boston

Beacon Press
25 Beacon Street
Boston, Massachusetts 02108-2892

Beacon Press books are published under the auspices of the
Unitarian Universalist Association of Congregations.

99 98 97 96 95 94 93 8 7 6 5 4 3 2

Text design by Christine Leonard Raquepaw

The authors gratefully acknowledge permission of the following:
International Museum of Photography at George Eastman House for Lewis W.
Hine's photograph, "A ten-year-old spinner, Cottonmill, North Carolina,
1909"; Earl Dotter for his photographs, "Cotton-bale opening room,
Watershoals, South Carolina" and "Autoworker mounting roofracks, assembly
line, Detroit, Michigan"; Nicole Hollander for her "Sylvia" comic strip; and
Newsweek, Inc. for the cover of *Newsweek,* February 24, 1992.

Library of Congress Cataloging-in-Publication Data

Hubbard, Ruth, 1924–
    Exploding the gene myth: how genetic information is produced and
manipulated by scientists, physicians, employers, insurance companies,
educators, and law enforcers / Ruth Hubbard and Elijah Wald.
      p. cm.
    Includes bibliographical references and index.
    ISBN 0-8070-0418-9 (cloth)
    ISBN 0-8070-0419-7 (paper)
    1. Medical genetics—Moral and ethical aspects. 2. Medical genetics—
Social aspects. I. Wald, Elijah. II. Title.
    [DNLM: 1. Ethics, Medical. 2. Genes. 3. Genetics, Medical.
QZ 50 H876e]
EB155.H8   1993
174'.9574—dc20
DNLM/DLC
for Library of Congress   92-48822
                                                        CIP

To Hella and Richard

who contributed much more than their genes

"WE USED TO THINK OUR FATE
WAS IN THE STARS.
NOW WE KNOW, IN LARGE MEASURE,
OUR FATE IS IN OUR GENES."

—JAMES WATSON, *Time,* March 20, 1989

"WE CANNOT THINK OF ANY SIGNIFICANT
HUMAN SOCIAL BEHAVIOR THAT IS
BUILT INTO OUR GENES IN SUCH A WAY
THAT IT CANNOT BE SHAPED
BY SOCIAL CONDITIONS."

—*Not in Our Genes: Biology, Ideology, and Human Nature,*
by R. C. LEWONTIN, STEVEN ROSE, AND LEON J. KAMIN

Contents

PREFACE: WHY THIS BOOK

A revolution is happening in the biosciences. Newspapers and magazines constantly report discoveries of genes for this or that disease, disability, or ability and many people believe that new biotechnologies will transform our lives more profoundly than transistors and computers have done. Yet genetics remains a specialized subject, and few people are equipped to evaluate how the new wonders will affect them. Words such as "genes" and "DNA" fly about. But what are genes and DNA, and how do they function?

We need to have a realistic sense of the positive contributions genetics and biotechnology can make, and of the risks inherent in the science, its applications, and its commercialization. We also need to understand that biotechnology can change not only how we live but how we think of ourselves and other animals. Are living organisms machines, so that it is safe to replace a gear here and a cog there, or are we too complex for anyone to foresee the effects of genetic tinkering?

This is a critical time in the development of genetics and biotechnology. Legislatures, courts, government agencies, and commissions are breaking new ground and making decisions about questions such as whether our genes can be patented or stored in data banks, how to prevent new forms of discrimination based on genetic information, and how to keep genetic information private.

It is crucial that we, as citizens, not leave this process in the hands of "experts." Like other people, scientists are interested in seeing their projects flourish, and their enthusiasm can blind them to the possible negative effects of their work. Since we will all have to live with those effects, we must become sufficiently informed to be able to decide to what extent genetics and biotechnology can improve our lives. We cannot just sit by as passive worshipers or victims.

This book is intended to provide an overview of what is occurring in modern genetics and to make it easier to understand and evaluate current applications of genetic research. In this rapidly changing field, in which scientific papers often are outdated even before they appear in print, it would be foolish for me to pretend to cover the most recent findings. Rather, I want to provide something like a basic survival handbook, a compass, and a few guideposts. For readers not trained in biology, I will try to present enough science to let you form your own opinions about the reports you read in the press. In addition, I want to give some historical insight into the destructive consequences of past overuses and misuses of genetics by scientists, physicians, and politicians, in this country and in Europe. At the end of the book, I have included a glossary of scientific terms and a list of books and organizations which can provide further information on the various subjects I discuss.

Although two of us have worked on this book, it is written in the first person. This is because I, Ruth Hubbard, am a biologist and take responsibility for the scientific content and for much of the interpretation we present. My coauthor, Elijah Wald, is a writer and a musician who believes, as I do, that anything worth saying can be said clearly enough so that people without special training can understand it.

I could not have written this book without the help of numerous friends and colleagues over many years. First and foremost I must mention my fellow members on the Board of Directors and the Human Genetics Committee of the Council for Responsible Genetics: Philip Bereano, Paul Billings, Liebe Cavalieri, Terri Goldberg, Colin Gracey, Mary Sue Henifin, Jonathan King, Sheldon Krimsky, Richard Lewontin, Abby Lippman, Karen Messing, Claire Nader, Stuart Newman, Judy Norsigian, Barbara Rosenberg, Marsha Saxton, Susan Wright, and Nachama Wilker, executive director of the Council for Responsible Genetics. Our collective work and discussions have guided and clarified my thinking about all the issues I discuss in this book. I have also profited from discussions with Charles Baron, Alice Daniel, John Roberts, Melvin Schorin, and Ernest Winsor, my fellow members of the Medical-Legal Committee of the Civil Liberties Union of Massachusetts. Over the years, I have profited from ongoing conversations with Rita Arditti, Adrienne Asch, Jon Beckwith, Joan Bertin, Lynda Birke, Robin Blatt, Carolyn Cohen, Richard Cone, Mike Fortun, Robin Gillespie, Stephen Jay Gould, Evelynn Hammonds, Donna Haraway, Sarah Jansen, Evelyn

Fox Keller, Renate Duelli Klein, Nancy Krieger, Suzannah Maclay, Emily Martin, Everett Mendelsohn, Laurie Nsiah-Jefferson, Cristian Orego, Rayna Rapp, Hilary and Steven Rose, Barbara Katz Rothman, Sala and Alan Steinbach, Nadine Taub, David Wald, Michael Wald, and many other colleagues and friends. I owe special thanks to Robin Gillespie, Mary Sue Henifin, Richard Kahn, Richard Lewontin, and Stuart Newman for critically reading parts or the entire text of this book, to Nancy Newman for preparing the index, and to Marya Van't Hul, our editor at Beacon Press, for many useful suggestions. I would like especially to acknowledge my husband George Wald's continuing interest and support during all the years since I became obsessed with this subject. All these people have helped me more than I can say and I thank them, but of course I alone am responsible for any errors.

ONE

. . . .

OF GENES AND PEOPLE

THE ROLE OF GENETICS IN OUR LIVES

We meet with genetics all the time, though we don't always recognize it. When we go to a doctor, we are first asked about our "family history," the diseases our parents or siblings have had. Only later, after the doctor has begun to form a theory about our problems, are we likely to be asked questions about our lives: where we live, what we eat, and the way we live in general. Despite the wide range of occupational health hazards, we are rarely asked about our jobs unless we have specific work-related complaints.

This "family history" is an attempt to come up with a genetic framework into which our problems can be fitted. The doctor uses this information about our relatives' health conditions as an aid in predicting what we may expect in our own lives. Such histories can include only what we happen to know about our family and therefore give only a rough picture. Modern genetic research tries to go beyond that, by looking at manifestations of inherited traits and eventually at genes themselves.

Such histories, whether based on family anecdotes or medical tests, are also looked at when we want to buy health or life insurance. They may determine whether we receive coverage and what premiums we will have to pay. More and more, they are required by prospective employers, and can affect whether or not we get a job.

A generation ago, people thought mostly about their economic and family situations when they considered whether they should have children. Today, they are often expected to undergo medical tests at every stage of the process, from premarital or preconceptive blood tests to amniocentesis during pregnancy. All this information is supposed to be useful. Doctors hope that it will give them a better understanding of our

1

health problems, and help them prevent or cure these problems. Insurers and employers hope that it will let them predict their future liabilities. We hope that it will help us to remain healthy and have healthy children.

The problem with linking all our health conditions to genes is that it makes us focus on what is happening inside us and draws our attention away from other factors that we should be considering. The genetic epidemiologist, Abby Lippman, has called this process *geneticization*. She writes:

> Geneticization refers to an ongoing process by which differences between individuals are reduced to their DNA codes, with most disorders, behaviors and physiological variations defined, at least in part, as genetic in origin. It refers as well to the process by which interventions employing genetic technologies are adopted to manage problems of health. Through this process, human biology is incorrectly equated with human genetics, implying that the latter acts alone to make us each the organism she or he is.[1]

Currently a new industry is being built on hopes of better living through genetics. Molecular biologists—the scientists who study the structure and function of genes and DNA—are acting as directors, consultants, and shareholders in biotechnology companies that seek to capitalize on every aspect of genetic research. Firms with names like Biogen, Genentech, Genzyme, Repligen, NeoRx, and ImClone are producing everything from predictive tests to drugs, hormones, and modified genes.

Biotechnology firms have been expensive to set up, and have lured investors who expect rich profits in the near future. This means not only that they have to put products on the market as soon as possible but they must create a market for those products. They are producing a host of tests and medications, and making glowing promises about the benefits to be derived from the use of these products. The evidence to support such promises is often slight or even nonexistent, but since most of the medical and scientific experts in the field are also connected with the industry they are inclined to be optimistic.

While the benefits of the new products are often illusory, disadvantages are appearing which are all too real. People have been refused employment or insurance on the basis of genetic tests whose results have no significance. Women have been needlessly frightened about the outcome of their pregnancies. Treatments with potentially harmful effects have been started without sufficient testing.

A more basic problem is that genetic tests and modifications encourage us to look upon ourselves as a collection of tiny discrete parts, rather than as whole human beings. Since we ourselves cannot do anything to change these parts, we are forced more and more frequently to entrust ourselves to specialists who supposedly can. Yet it often makes much more sense to deal with the whole human being, rather than to tinker with the parts. This can involve things that are much more in our control, such as changing where or with whom we live, what and how much we eat, or other aspects of our lives.

The process of reducing objects or organisms to their smallest parts rather than looking at them as a whole is called *reductionism,* and it is not confined to genetics. In the last century or so, reductionism has become a major force in science. From Pasteur's bacteria to the physicist's atoms, we have grown used to the idea that the smallest things can have the most overwhelming effects. In biology, reductionism fosters the belief that the behavior of an organism or a tissue can best be explained by studying its cells, molecules, and atoms and describing their constitution and function as accurately as possible. However, reductionists often lose sight of the forest in their zeal to examine the ridges on the twigs of the trees.

In the biological sciences, the status once enjoyed by naturalists, who observe how animals live and what they do, has shifted to molecular biologists, who study DNA molecules and segments of these molecules which they call genes. Most modern biologists believe that work at the molecular level will yield a more profound understanding of nature than they could get from the study of cells, organs, or entire organisms. The fact that experiments with animals are more difficult to control or duplicate than experiments in test tubes has made it easy to dismiss the former as fuzzy science. Molecular biology has therefore become the most prestigious of the biological disciplines.

In the last few years, molecular biologists under the auspices of the National Institutes of Health and the Department of Energy have mounted a project that has been compared to the U.S. space program for its scope and expense (a projected three billion dollars in fifteen years). Called the Human Genome Project, it is an attempt to map and sequence all the DNA of a human "prototype." This is reductionism at its most extreme, as genome scientists will be constructing a hypothetical sequence of submicroscopic pieces of DNA molecules, and will then declare that sequence to be the essence of humanity.

Harvard molecular biologist and Nobel Laureate, Walter Gilbert, has referred to the human genome as the "Holy Grail" of genetics.[2] Such imagery, intended to elicit a religious awe for the wonders of science, has become common among genome scientists and is carried over into most media reports on the project. For instance, a "NOVA" television program on the human genome referred to it as the "Book of Life." James Watson, author of *The Double Helix* and the first director of the National

3

Center for Human Genome Research, avoids explicit religious metaphors, but says his objective is "the understanding of human beings" and of life itself.[3]

Being human, however, is not simply a matter of having a certain DNA sequence. Molecular biologists are no more qualified than the fabled guru on the mountaintop when it comes to telling us the meaning of life. They can give some answers for some aspects of the question, but their answers are useful only in specific contexts.

GENES FOR DEAFNESS, GENES FOR BEING RAPED

While most people have never heard of the Genome Project, no one can miss the flood of gene stories in the popular press. For instance, one day's "Medical Notebook" section in the *Boston Globe* contained these four headlines: "Genetic link hinted in smoking cancers." "Schizophrenia gene remains elusive." "A gene that causes pure deafness found." "Do the depressed bring on problems?"[4]

Once, this emphasis on genes would have seemed surprising, but in the last few years such stories have become commonplace. We all see the articles, but we do not always bother to read them. For most of us, genetics remains something complicated, scientific, and a bit boring. And yet, the subjects being discussed are often very close to home. They include alcoholism, cancer, learning problems, mental illness, sex differences, and such basic processes as aging. —

The four *Globe* stories are typical of most current reporting on genetics, both in the mass media and in scientific journals. They contain a mix of interesting facts, unsupported conjectures, and wild exaggerations of the importance of genes in our lives. A striking thing about much of this writing is its vagueness. In the first headline, for example, a "link" to smoking cancers is "hinted." The story itself says "a report released this week . . . *suggests* certain individuals *may* carry a gene that makes them especially vulnerable to smoking-related cancers" (italics mine). It then tells us that the researcher estimates that 52 percent of the population "may" have such a gene, "if it exists." In other words, it is possible that slightly more than half of us are particularly susceptible to lung cancer if we smoke. The remaining 48 percent of us may perhaps be less susceptible, although smokers are still at significantly greater risk than nonsmokers.

Even if such a "cancer gene" were isolated, that would not change the fact that smoking is harmful, nor would it help people to quit smoking or doctors to treat cancer. This information would therefore not be useful to most newspaper readers, even if the article contained valid scientific conclusions. So why is it published? One reason is that both genes and the dangers of smoking currently are of interest to a lot of people. Another is that such information could be extremely useful to cigarette companies. As people with lung cancer are beginning to sue these companies, the companies would love to be able to blame the cancers on these people's genetic "susceptibilities." If the people bringing suit turned out to be in a special high-risk group, the companies could disclaim responsibility. If that high-risk group includes over half the population, that is not the companies' problem.

Many new genetic breakthroughs are like this. They do not make people healthier; they merely blame genes for conditions that have traditionally been thought to have societal, environmental, or psychological causes. News reports about such studies fuel the widely held perception that our health problems originate inside us and draw attention away from outside factors that need to be addressed. Scientists did not create this perception, but they contribute to it when their interest in genes keeps them from emphasizing, or even admitting, that there are other ways to explain our health problems.

Witness the next piece in the "Medical Notebook." It begins, "A series of attempts to confirm the existence of a gene for schizophrenia have failed, three years after the announcement of an apparent gene link caused a stir among mental health researchers." If a link cannot be confirmed after repeated attempts, that would seem to suggest that the condition is not genetic. However, the column quotes a psychologist named Irving Gottesman as saying that "studies continue to indicate that a gene or genes creates 'risk-enhancing factors' for schizophrenia."

The studies he refers to show that people who have schizophrenic siblings are somewhat more likely to be schizophrenic than people who don't. Since many psychiatrists think that schizophrenia is caused by family problems, this result is not at all surprising. To call it "evidence" of genetic factors is at the very least misleading.

Like the smoking study, this is a story built on air. Such articles suggest that genes are involved in all sorts of conditions and behaviors, but all they really tell us is that a lot of money is being spent on genetic research. The grandiose nature of the claims disguises the fact that the research is not particularly newsworthy.

The next *Globe* story gives an example of the more responsible kind of genetic research. Scientists have identified "a gene that causes pure deafness," the first such gene to be found. All the people with this particular form of deafness are members of one extended family in Costa Rica. The fact that this gene has been isolated may help scientists to understand other kinds of deafness as well, though that remains to be seen.

This sort of basic research increases our body of knowledge and can be useful. However, it is not the sort of discovery that normally gets into the daily papers. It is in the *Globe* because it is a gene story and, unpretentious as it is, adds solid facts to the featherweight claims in the other stories.

The myth of the all-powerful gene is based on flawed science that discounts the environmental context in which we and our genes exist. It has many dangers, as it can lead to genetic discrimination and hazardous medical manipulations. The last *Globe* piece is an extreme example of the dangerous and unwarranted conclusions that are sometimes drawn from genetic research. It reports a survey by Lincoln Eaves, a behavioral geneticist, of research by various investigators on twelve hundred pairs of female twins whom the investigators considered to be prone to depression. Eaves said he found evidence of genetic causes for this depression, though the evidence is not provided in the article.

Eaves also administered a questionnaire "asking whether the volunteers had suffered traumatic events, such as rape, assault, being fired from a job, and so forth." He found that the women who were chronically depressed had suffered more traumatic events than those who weren't.

Now, if he were not assuming that their depression was genetic, he might suspect that they were depressed because of the bad things that had happened to them. However, his interest is genetics. So, the article continues, Eaves "suggested that [the women's] depressive outlook and manner may have made such random troubles more likely to happen."

What kind of reasoning is that? The women had been raped, assaulted, or fired from their jobs, and they were depressed. The more traumatic events they had experienced, the more chronic the depression. This suggests that depression brings on problems? If Dr. Eaves had found that football players frequently get fractures, would he have suggested that brittle bones make people play football? It might have been worth looking for a genetic link if he had found that the depression was not related to any life experience. But once he found a clear correlation between traumatic events and depression, why look for a genetic explanation?

Ridiculous as this research may be, the press reports it with a straight face. At present, genes are newsworthy and virtually any theorizing about them is taken seriously. This is not the fault of the media. Science, government, and business are all hailing genetics and biotechnology as the wave of the future.

A WORD ABOUT SCIENTISTS

I want to emphasize at the outset that, contrary to popular belief, scientists are not detached observers of nature and the facts they discover are not simply inherent in the natural phenomena they observe. Scientists construct facts by constantly making decisions about what they will consider significant, what experiments they should pursue, and how they will describe their observations. These choices are not merely individual or idiosyncratic but reflect the society in which the scientists live and work.

For example, it should not surprise us that nineteenth-century biologists, who were by definition male, found scientific reasons why girls could, or should, not get the same education as boys. Some of them "proved" that women's brains were smaller than men's, others that education damaged girls' reproductive organs so that educated women would not be able to have children. It is easy to recognize now that these scientific descriptions grew out of the beliefs of the time in which the scientists were living. But today scientists are still trying to explain differences in the language, spatial, and mathematical abilities of girls and boys in terms of brain structure and genes. It is more difficult for us to identify the various ideological roots of our contemporary science, since all of us share these roots to some extent.

Scientific education initiates students into a cultural enterprise, with its own history and system of beliefs. One of those beliefs is that the march of science is immune from political and societal pressures, that scientists can function in an ideological vacuum. This belief has been proved wrong time and again. Scientists, as a group, tend to provide results that support the basic values of their society. This is not surprising, since scientists live in that society and make their observations with that society's eyes.

This is particularly true when scientists are studying people. In the case of human genetics and molecular biology, we must expect the value our society attaches to genealogy and heredity to influence every stage of research and discussion. Societal values are automatically coded into the scientific meanings of terms like "inherited traits" and "genes." This is not to say that scientists deliberately misrepresent what is happening in nature or that their descriptions are necessarily wrong. But, especially in areas that touch closely on cherished beliefs about ourselves, about our society, and about other living beings—and genetics does all of that—science is likely to reflect those beliefs.

DNA, the molecule, is material and real and may well have the physical structure described in biology textbooks. However, our understanding of DNA and genes incorporates ideological baggage derived from our concepts of health and disease, normality and deviance, and what we can be or ought to be. It would be good if scientists took this into account, but, unfortunately, scientific training tends to obscure connections between science and society. Students are taught that science starts with basic questions that scientists try to answer, and develops as the answers lead to further questions. Most scientists believe this. They ignore the ongoing interplay of science with society or, if they consider it, emphasize the effects of science on society rather than the ways in which society affects scientists' perceptions and preconceptions.

To understand science in its social context, we must always be conscious of the interactions among scientific practices, descriptions, and interpretations and the cultural beliefs and economic circumstances within which scientists operate. Otherwise we cannot understand the way scientific facts are created and the way these facts are put to use in society at large.

I want to make it very clear that I do not think molecular biologists and other genetic researchers are doing their science badly. They are simply doing science-as-usual, with the narrow focus inherent in that approach.

An analogy may clarify what I mean. For centuries, historians looked at the past in terms of the accomplishments of great men (specifically, great *white* men). The history most of us learned in school was constructed around kings and generals and such. In this century, there has been a gradual shift from this "great man" theory of history to social history, the attempt to understand the past by finding out what we can about how ordinary women and men lived. So, historians have moved from a model in which history was explained in terms of the activities of "important men" to a more integrative one in which they try to get as complete and varied a picture as possible.

Molecular biology has taken the opposite path, moving from an integrative biology to a biology that traces everything back to the "great molecule"—DNA. My disagreement is with this general view. Only rarely do I find fault with the way specific experiments are done or interpreted. What I object to is the reductionist effort to explain living organisms in terms of the workings of "important molecules" and their component parts.

HEREDITY AND ENVIRONMENT

Genetic research tries to answer a set of questions that have probably been around for as long as there have been people to ask them. These questions deal with the interplay between heredity and environment and between similarity and difference.

No one would confuse me with my daughter, but there are clear similarities between us. As humans, each of us is unique, but we are clearly one species, far more like one another than we are like chimpanzees or sea gulls. While many of the differences between us arise from our environment, others obviously are products of heredity.

Throughout Western intellectual history, people have argued about which is most significant: heredity or environment, nature or nurture. Belief in the power of heredity was strong at the beginning of this century. Genetics was becoming a discipline in its own right, and scientists hoped it would solve a host of problems. However, hereditarian ideas went out of fashion after the Second World War. When the disastrous consequences of the Nazis' racist, hereditarian policies became widely known, they stood as a horrific warning of the dangers of assigning too much power to biological inheritance and using genetic means to improve humanity.

Since the early 1970s, the pendulum has been swinging back and scientists are again emphasizing the importance of heredity in shaping our character and actions. This shift is due in part to a conservative backlash against the gains of the civil rights and women's rights movements. These and similar movements have emphasized the importance of our environment in shaping who we are, insisting that women, African Americans, and other kinds of people have an inferior status in American society because of prejudices against them, not because of any natural inferiority. Conservatives are quick to hail scientific discoveries that seem to show innate differences which they can use to explain the current social order.

Like reductionists, hereditarians try to find simple answers to complicated questions. But the interactions and transformations that go on inside us and between us and our environment are too complex to be forced into such simplistic patterns. Our environment is full of other living organisms, from the bacteria that colonize our intestines and supply us with essential vitamins and other foodstuffs to the human beings and other animals with whom we live. Looking at all our genes, or even at all the genes of all these creatures, would still not tell us very much about our interrelationships in societies and in nature.

True, once a woman's egg and a man's sperm have fused, the genes of a future person have come together, but many things that happen in the fertilized egg, in the womb, and after the person is born will also have major effects. If the pregnant woman cannot get proper nourishment or receives inappropriate medication, the embryo may be damaged or even die. Conversely, if she has the opportunity to take proper care of herself, and of the child once it is born, this will have positive effects on the child's development. Like its genes, these circumstances will be an integral part of that child. Infants must be nurtured into human beings. Left entirely to their own devices, they are like the boy raised by wolves—technically human, but unable to engage in even the most basic human interactions.

As we all know, people who grow up in different cultures can be remarkably different, not only in personality but also in physique. Within a generation, and without any genetic change, immigrants from Asia and southern Europe have seen their American-born children tower over them. Even living within one family, each of us is subject to different influences that can affect us both mentally and physically.

External influences act on us throughout our lives. I remember a student of mine who suffered an acute depression as a result of kidney failure brought on by an antibiotic. When I visited him in the hospital, I could not recognize him. In the span of a week, this athletic young fellow had turned into a hollow-cheeked old man, with an old man's voice and gait. Fortunately, these changes turned out to be reversible, but they were a striking example of how quickly we can experience profound biological changes and become someone different and unexpected.

Stupid people can become bright, apathetic people alert, or decrepit people lively and strong as a result of changes in their life circumstances. Many social movements, from feminism to civil rights struggles, have been based on the possibility of making such changes, and have succeeded in producing "miraculous" results.

Still, genetic explanations have their attractions. We have just gone through a period when parents' actions were blamed for whatever went wrong with their children. It should not surprise us that parents who have had to walk a thin line between being "overly permissive" and "overly strict," being too "distant" or "smothering" their children with affection, would embrace the idea that genes, rather than their own mistakes, are at the root of their children's problems. But genetic explanations are as confining as they are liberating. Genes participate in all the ways we function, but they do not determine who we are. They must affect our development, but so do a host of personal and social circumstances.

So, what part do genes in fact play in our lives? We do not know the answer, and we cannot ever hope to know it. Humans, or even fruit flies, are complex organisms leading complex lives, and our experiences and our biology interact in unpredictable ways. Neither genetics nor molecular biology can tell us all that much about people. They can only tell us about our genes.

WHAT ARE GENES?

But what are genes? Different kinds of biologists have answered that question in different ways. To molecular biologists, a gene is a stretch of DNA that specifies the composition of a protein and may affect whether and at what rate that protein is synthesized, as well as sometimes affecting the synthesis of proteins specified by nearby genes. To geneticists, genes are parts of our chromosomes that mediate heritable characteristics or traits. To population biologists, genes are units of difference that can be used to distinguish various members of a population from each other. To evolutionary biologists, genes are historical records of the changes organisms have undergone over time. All these definitions overlap and complement each other, and which one a particular scientist focuses on simply depends on her or his interest.

Biologists agree that genes are functional segments of DNA molecules, but the word "gene" predates that definition. The term was invented at the beginning of this century to denote particles that were thought to mediate the expression of hereditary traits in individuals and to transmit these traits from parents to their offspring. Later it became clear that there were no such particles, but that these functions are performed by portions of DNA molecules.

In this book, I will use the terms "gene" and "DNA segment" interchangeably. "DNA segment" or "functional DNA segment" is more accurate, but "gene" remains a useful shorthand. However, since this word has come to have an almost iconic power, we should remember that it is only a simplification of a complex reality.

The language that geneticists use often carries considerable ideological baggage. Molecular biologists, as well as the press, use verbs like "control," "program," or "determine" when speaking about what genes or DNA do. These are all inappropriate because they assign far too active a role to DNA. The fact is that DNA doesn't "do" anything; it is a remarkably inert molecule. It just sits in our cells and waits for other molecules to interact with it.

In a way, the DNA in our cells is like a cookbook. We need a cookbook if we want to make a complex dish, but it does not make the dish, nor can it determine which dish to make or whether the dish will come out right. The cook and the ingredients will determine whether and how a recipe is used, whether we end up eating soup or cake, and how the food tastes. Cookery is also an apt metaphor because it introduces an element of adaptability and flexibility. A good cook can deviate from the recipe and fudge the outcome if she or he lacks some seemingly essential ingredients or implements. Similarly, cells and organisms can compensate for "genetic mistakes." Moreover, if cells and organisms are the cooks in this metaphor, many ingredients, among them genes and environmental factors, combine to produce a "dish" that could not have been predicted by looking at the ingredients separately.

All too often genes are looked on as absolute predictors. When people speak of genes "for" this or that trait, they convey an aura of inevitability which limits us. The belief that our capacities are encoded in our genes can prevent us from taking available steps to change ourselves or the conditions of our lives. We must remember that genetic functions are embedded in complex networks of biological reactions and social and economic relationships. They are not simple processes that can be duplicated in a laboratory. Also, because of the importance of genes to our image of ourselves, scientists find it difficult to look at them objectively.

In the following chapters, I shall explore the ways scientific knowledge about genetics is both framed by and affects our society's beliefs about normalcy, and about health and disease. A good way to start is by looking at eugenics and at the modern efforts to improve the biological constitution of human beings by means of genetic selection and manipulation.

TWO

· ·

GENETIC LABELING
AND THE OLD EUGENICS

THE BIRTH OF EUGENICS

A desire to understand the present and foresee the future has existed in virtually all cultures. People have consulted shamans, oracles, priests, witches, or astrologers, whose diagnostic tools range from direct heavenly communications to the interpretation of dreams, tea leaves, or the movements of the stars.

Most such soothsayers spend their lives supporting the status quo. A prophet who promises the wealthy that their future will be miserable is not going to have a pleasant or lucrative career. One who harps on social injustices or the virtues of the poor is considered a fool or a dangerous agitator. Every royal court has had its magicians and prophets, and they have spent most of their time telling the king that he is singularly well favored by the stars and will go down in history as an exemplary monarch.

With the rise of the merchant classes and the scientific renaissance, the power of kings and magicians went into decline, and the industrial revolution put them virtually out of business. The early industrial barons believed in reason and cold, hard facts. Yet, like the kings before them, they were not interested in hearing bad news. Now, science was used to explain that those on top of the heap were there because of their innate superiority to the masses, who were on the bottom because they were of inferior stock.

A fine example of using of science to explain social status can be seen in the following paragraph from *Heredity in Relation to Eugenics,* published in 1913 by Charles Benedict Davenport, a professor of biology at Harvard and later at the University of Chicago:

Pauperism is a result of complex causes. On one side it is mainly environmental in origin as, for instance, in the case when a sudden accident, like death of the father, leaves a widow or family of children without means of livelihood, or a prolonged disease of the wage earner exhausts savings. But it is easy to see that in these cases heredity also plays a part; for the effective worker will be able to save enough money to care for his family in case of accident; and the man of strong stock will not suffer from prolonged disease. Barring a few highly exceptional conditions poverty means relative inefficiency and this in turn usually means mental inferiority.[1]

Hereditarianism produced beautifully self-fulfilling prophecies. Anyone who succeeded was, ipso facto, a superior person. Since the children of the wealthy and educated usually turn out to be wealthy and educated, while the children of the poor tend to remain poor, it was quite clear to hereditarians that talent ran in the family. As for the social and natural scientists who produced the body of hereditarian theory, they were not only demonstrating the worthiness of their patrons but also proving their own superiority over the backward peoples who had failed to create the wonders of modern science.

The term *eugenics*, which means "wellborn," was coined in 1883 by Francis Galton, who came from a distinguished British upper-class family and was a cousin of Charles Darwin. Galton wrote that he invented it in order to have "a brief word to express the science of improving the stock, which is by no means confined to questions of judicious mating, but which, especially in the case of man, takes cognizance of all the influences that tend in however remote a degree to give the more suitable races or strains of blood a better chance of prevailing speedily over the less suitable than they otherwise would have had."[2] Given that Galton had little doubt about who represented "the more suitable races or strains of blood," class and race biases were there from the start of the eugenics movement. Galton later helped to found the English Eugenics Society and became its honorary president.

Many people have come to associate eugenic thinking with political conservatism because the early British and American eugenicists tended to be conservative and incorporated their class, race, and imperialist biases into the scientific and political programs of the eugenics movement. But, particularly when we come to look at contemporary hereditarian and eugenic thinking in the next chapter, it is important to remember that support for eugenics has come from across the political spectrum. In the nineteenth century, not only conservative followers of the Reverend Thomas Malthus but also progressives and liberal believers in meritocracy believed that eugenics held promise for human betterment.

In 1912, near the start of a long career as a geneticist, Hermann J. Muller wrote: "The intrinsic interest of these questions [about heredity] is matched by their extrinsic importance, for their solution would help us to predict the characteristics of offspring yet unborn and would ultimately enable us to modify the nature of future generations."[3] Muller was a politically progressive idealist, who tried to emigrate to the Soviet Union in the early 1930s because he wanted to help build a better world.

Until World War II, many distinguished biologists and social scientists in Great Britain and the United States supported eugenics, or at least did not express their opposition to it. As late as 1941, when eugenic extermination practices were in full force in Nazi Germany, the distinguished British biologist Julian Huxley—brother of Aldous Huxley, the author of *Brave New World*—wrote an article called "The Vital Importance of Eugenics," which begins: "Eugenics is running the usual course of many new ideas. It has ceased to be regarded as a fad, is now receiving serious study, and in the near future will be regarded as an urgent practical problem."[4] Later in the article, he argued that society must "ensure that mental defectives shall not have children." In a not unusual blurring of eugenic and economic considerations, he defined as mentally defective "someone with such a feeble mind that he cannot support himself or look after himself unaided."

Huxley hesitated to prescribe whether "racial degeneration" should be counteracted by "prohibition of marriage" or by "segregation in institutions" combined with "sterilization for those who are at large," but he stated as though it were an established fact that most "mental defects" are hereditary. Actually, though most instances of mental retardation among the middle and upper classes have a genetic component, this is not the case among poor people, where inadequate nutrition and prenatal care, lead poisoning, and substandard school systems play a considerable part.[5]

Like Muller, Huxley did not limit his concern to those persons who were demonstrably afflicted with "mental defects." He looked forward to a future when it would become possible "to diagnose the carriers of the defect [who are] apparently normal," since "if these could but be detected, and then discouraged or prevented from reproducing, mental defects could very speedily be reduced to negligible proportions among our population."

It is shocking to realize that at the very moment when the Nazis were sterilizing and killing adults and children who had been diagnosed as disabled or mentally ill, Huxley could voice regret that it was "very difficult to envisage methods for putting even a limited constructive program [of eugenics] into effect . . . due as much to difficulties in our present socioeconomic organization as to our ignorance of human heredity, and most of all to the absence of a eugenic sense in the public at large."

Huxley was by no means alone. Eugenics societies on both sides of the Atlantic were organizing "eugenic fairs" to educate the public about the menace of inherited defects, and warning the upper classes about the dangers of "class suicide" because they were having too few children while poor people were having too many.

European eugenicists tended to be preoccupied with class differences, but in the United States ethnic and racial concerns were at the top of the agenda. Lewis Terman, one of the principal engineers and advocates of IQ testing, expressed some of these in an article he published in 1924, in which he worried that

> the fecundity of the family stocks from which our most gifted children come appears to be definitely on the wane. . . . It has been figured that if the present differential birth rate continues 1,000 Harvard graduates will, at the end of 200 years, have but 56 descendants, while in the same period, 1000 South Italians will have multiplied to 100,000.[6]

GENETIC LABELING

To formulate eugenic policies, it is necessary first to label certain physical or mental traits and social behaviors as aberrant and then to assume that they are transmitted biologically from parents to their children. Such labels can easily be exploited for ideological and political ends. An extreme example is the invention of *drapetomania,* a hereditary mental disease said to be prevalent among black slaves in the South, which manifested itself in an irresistible urge to run away from their masters.

There was no evidence whatsoever for the biological nature or genetic transmission of many of the traits that eugenicists said were inherited. Take the label "feebleminded." Class, race, ethnic, and linguistic differences have often led to people's, and especially children's, mental capacities being judged lower than average. Many instances of slow or arrested mental development have their origins in infectious diseases or physical, psychological, or social traumas and hence are not inherited biologically, though members of the same family may share them because they share these experiences.

When we come to such labels as "alcoholism" or "pauperism," difficulties of assigning causes are compounded by problems of description and definition. How much does one have to drink to be an alcoholic? How poor does one have to be, and for how long, before one is a pauper?

Even where diagnosis is relatively unambiguous, the eugenic approach only confuses and obscures the issues. In the early part of this century, *pellagra*, a chronic condition characterized by skin eruptions, digestive and nervous disturbances, and eventual mental deterioration, reached epidemic proportions in parts of the southern United States. Many people believed it to be infectious in origin, like syphilis. Charles Davenport and his colleagues agreed, but argued that by virtue of genetic predisposition pellagra is acquired preferentially by certain kinds of people. With uncanny precision, Davenport specified that "when both parents are susceptible to the disease, at least 40 percent, probably not far from 50 percent, of their children are susceptible."[7]

Around the same time Joseph Goldberger, an American epidemiologist, showed that pellagra results from a lack of a vitamin found in grains and fresh vegetables, which he called the *pellagra preventive* (or *PP*) *factor*. His PP-factor was later renamed nicotinamide or niacin and identified as a member of the vitamin B complex. Though Goldberger published his findings in 1916, nutrition programs that could have prevented many cases of pellagra were not instituted until the beginning of the New Deal, in 1933. The conservative Republican administrations of the 1920s, averse to spending money on health and nutrition, were supported in their inertia by the assertions of Davenport's adherents that pellagra was hereditary and could not be remedied by social programs.

The old eugenics reached its ultimate extreme in the Nazi extermination programs. Initially, these were directed against the same sorts of people eugenicists had targeted in Great Britain and the United States—people labeled as having physical or mental disabilities. Later, these programs were expanded to include Jews, homosexuals, Gypsies, Eastern European "Slavs," and other "inferior" types.

The Nazis referred to their procedures as "selection and eradication," to emphasize that they were doing nothing random or haphazard. They were proud of the fact that the exterminations were scientifically planned and carried out and that, even in the death camps, "eradication" was always preceded by "selection."

The theoretical underpinnings of the Nazi extermination programs, and the enthusiastic sponsorship and support they received from German geneticists, anthropologists, psychiatrists, and other members of the so-called healing professions, have been described in grisly detail by Stephan Chorover,[8] Robert J. Lifton,[9] Benno Müller-Hill,[10] and Robert Proctor.[11] It is important to realize that the Nazis drew directly on eugenic arguments and programs developed by scientists and politicians in Great Britain and the United States. They just made these policies more inclusive and implemented them more decisively than British and American eugenicists may have intended.

In the United States, Charles Davenport was one of the most active supporters of eugenics. He persuaded the Carnegie Institution of Washington, D.C. to set up the Station for the Experimental Study of Evolution at Cold Spring Harbor on Long Island and became its first director in 1904. Part of the station's mission was to organize a central data bank in which to store genetic information on large numbers of individuals. In 1910, with funding from John D. Rockefeller, Jr. and from Mrs. E. H. Harriman, an heiress to the Harriman railroad fortune, Davenport opened the Eugenics Record Office at Cold Spring Harbor. He appointed Harry W. Laughlin, a Princeton Ph.D., as its superintendent, and recruited graduates of Radcliffe, Vassar, and the Ivy League schools to go and interview large numbers of so-called mental and social defectives.

Trained for a few weeks, in courses offered at Cold Spring Harbor and at a comparable institution administered by the psychologist Henry H. Goddard at Vineland, New Jersey, these upper-class field-workers descended on poor communities in New York and New Jersey. Despite their scant training, and the class and ethnic differences between themselves and their subjects, the field-workers were considered competent to diagnose, by sight alone, such varied "hereditary" conditions as "dementia," "shiftlessness," "criminalism," and "feeblemindedness."[12] After identifying a range of "mental defects," they turned these "data" into pedigree charts and scientific reports.

Despite the apparent vagueness of the terminology, Davenport and his associates felt that they could categorize these conditions with mathematical precision. For instance, with reference to "shiftlessness," Davenport wrote:

> Let us take "shiftlessness" as an important element in poverty. Then classifying all persons in . . . two families as very shiftless, somewhat shiftless, and industrious the following conclusions are reached. When both parents are *very* shiftless practically all children are "very shiftless" or "somewhat shiftless." Out of 62 offspring, 3 are . . . "industrious" or about 5 per cent. When both parents are shiftless in some degree about 15 per cent of the known offspring are recorded as industrious. When one parent is more or less shiftless while the other is industrious only about 10 per cent of the children are "very shiftless." It is probable that . . . shiftlessness . . . [is] due to the absence of something which can be got back into the offspring only by mating with industry.[13]

A ten-year-old spinner, Cottonmill, North Carolina, 1909. (Photograph by Lewis W. Hine, © by International Museum of Photography at George Eastman House.)

Despite the shoddiness of this science, Davenport's Eugenics Record Office and its staff were major resources for the two legislative programs that became cornerstones of U.S. eugenic policies: the involuntary sterilization laws and the Immigration Restriction Act of 1924.

INVOLUNTARY STERILIZATION

Following Galton's original suggestions, eugenics programs were of two kinds, positive and negative. Positive eugenics was intended to encourage the "fit" (read: healthy, successful, well-to-do) to have many children. Negative eugenics was meant to prevent the "unfit" (an extremely elastic category) from having any.

Laughlin and Davenport favored the sterilization of people they referred to as "hereditary paupers, criminals, feebleminded, tuberculous, shiftless and ne'er-do-wells,"[14] but they did not originate the notion. As early as 1897, the Michigan legislature considered—and defeated—a bill to sterilize people of "bad heredity." A subsequent, compulsory sterilization bill, aimed at "idiots and imbecile children," was passed by the Pennsylvania legislature in 1905, but vetoed by Governor Samuel Pennypacker with the strong message that it was not only "illogical" but "violates the principles of ethics." [15] The first compulsory sterilization law that was actually enacted passed the Indiana legislature in 1907. But even before that, in 1899, a Dr. Harry Sharp began to perform involuntary vasectomies at the Indiana State Reformatory at Jeffersonville on inmates whom he judged to be "hereditary criminals" or "otherwise genetically defective." [16]

The British geneticist J. B. S. Haldane, in his *Heredity and Politics*, gives an example of the standards used to make such judicial determinations. He quotes a Judge G. B. Holden of the superior court in Yakima County, Washington, as follows:

> On January 30, 1922, John Hill pleaded guilty to the crime of grand larceny. The theft was of a number of hams, which he took by stealth because of his impoverished condition. . . . Hill is a Russian beet sugar laborer, with a wife, and five children all under age of eleven years. He is robust physically, about forty years of age, and his wife some years his junior. Hill, his wife and five children are all mentally subnormal, even for their situation in life. . . . It was apparent that he could not provide them with the common necessities of life. . . . He was forced to steal to prevent them from starvation, or to apply for public aid. The case was brought to the attention of the authorities through the discovery of the theft of the hams, since which time he and his family are partially dependent upon public charity . . . ; with more children the extent of demand for public charity will be increased. [17]

Judge Holden found Hill guilty of grand larceny and gave him an indeterminate sentence of "not less than six months, nor more than fifteen years, imprisonment in the state penitentiary," to be suspended if he agreed to be sterilized. "Under these conditions," continues Judge Holden, "the operation was suggested to him, and after explanation . . . he consented." [18]

Haldane remarks that Judge Holden did not say what tests he used to determine that Hill and his family were mentally defective, and goes on to cite another of the judge's statements, this one about a burglar named Chris McCauley, whom he sentenced to compulsory sterilization: "This man, about thirty-five years of age, is subnormal mentally and has every appearance and indication of immorality. He has a strain of Negro blood in his veins, and has a disgusting and lustful appearance."[19]

Haldane summarizes the situation by suggesting that "Hill would not have been sterilized had he possessed an independent income," nor would McCauley have been, "had his complexion been lighter and his appearance more in conformity with Judge Holden's aesthetic standards."[20]

By 1931, some thirty states had compulsory sterilization laws on their books, aimed mostly at the "insane" and "feebleminded." These categories were loosely defined to include many recent immigrants and others who were functionally illiterate or knew little or no English and who therefore did poorly on IQ tests. The laws also were often extended to so-called sexual perverts, drug fiends, drunkards, epileptics, and others deemed ill or degenerate. Although most of these laws were not enforced, by January 1935 some twenty thousand people in the United States had been forcibly sterilized, most of them in California. The California law was not repealed until 1979 and, according to Phillip Reilly, a physician and attorney, in 1985 "at least nineteen states had laws that permitted the sterilization of mentally retarded persons (Arkansas, Colorado, Connecticut, Delaware, Georgia, Idaho, Kentucky, Maine, Minnesota, Mississippi, Montana, North Carolina, Oklahoma, Oregon, South Carolina, Utah, Vermont, Virginia, and West Virginia)."[21]

EUGENIC IMMIGRATION POLICIES

While the eugenicists diagnosed individuals with "hereditary defects" in virtually all ethnic groups, they found that certain groups had a much higher proportion of "defectives" than others. For this reason, eugenics was an explicit factor in the Immigration Restriction Act of 1924. This act was designed to decrease immigration to the United States from southern and eastern Europe, so as to tilt the population balance in favor of U.S. residents of British and northern European descent. The number of people allowed to immigrate into the United States from any country each year was restricted to 2 percent of the U.S. residents who had been born in that country, as listed in the census of 1890.

That thirty-four-year-old census date was chosen deliberately because it set as a baseline the composition of the U.S. population prior to the major immigrations from southern and eastern Europe at the end of the nineteenth century. In the 1930s and early 1940s, this legislation prevented the immigration of countless Jews who were attempting to flee the Nazis, because they had been born in eastern Europe.

Harry Laughlin of the Eugenics Record Office was one of the most important lobbyists and witnesses in favor of the Immigration Restriction Act at the congressional hearings that preceded its passage and was dignified with the title of "expert eugenical agent" by the House Committee on Immigration and Naturalization.

THREE

· ·

THE NEW EUGENICS: TESTING, SCREENING, AND CHOICE

OVERT AND SUBTLE EUGENICS

Interest in eugenics declined after the Second World War. Traditional co-lonialism was in retreat and the United Nations held out the hope of a future in which the peoples of the world could meet on an equal basis. Revulsion against Nazi eugenic practices led to a reaction against the whole idea of "better" and "worse" races. While racism did not disap-pear, people couched their racist attacks in different language, speaking of their targets as "underdeveloped," rather than genetically inferior. The suggestion was that we are all human, though some of us may be more culturally or socially advanced.

As eugenics became politically unacceptable, within the scientific community it was also losing favor for pragmatic reasons. Scientists be-gan to realize that most inherited conditions are *recessive*, rather than *dominant*. Someone with a dominant genetic condition will pass it on to roughly half of her or his descendants. But to inherit a recessive condi-tion, people must receive copies of the relevant *allele*, or form of the gene, from both their parents. If they inherit only one copy from one parent, they generally show no symptoms and are said to be *carriers* for that condition. Even if two carriers have children together, each child has only one chance in four of manifesting the condition.

Familiar examples of recessive conditions are *phenylketonuria*, or *PKU* (a metabolic problem that can result in mental retardation), *cystic fibrosis* (a glandular disturbance that leads to the accumulation of mucus in the lungs and to repeated infections), *Tay Sachs* disease (a fatal neurological disease of young children), and *sickle-cell anemia* (a blood disease that can be extremely painful and disabling).

Because these conditions are recessive, people who manifest them

23

(those with two copies of the affected allele) are only a fraction of those who carry at least one copy. This is because most alleles associated with recessive conditions are carried by people who have no symptoms and often have no reason to suspect that they are carriers. Recessive mutations are propagated by healthy, "normal" members of the population.

Early in this century, the British mathematician G. H. Hardy and the German physician W. Weinberg, working independently, developed a mathematical theorem for calculating the number of carriers for a recessive condition in a population by looking at the number of people who manifest the condition. For example, one in 25,000 people in the United States has PKU (that is, two copies of the relevant allele). Using the Hardy-Weinberg theorem, one can calculate from this that about one in eighty people has one affected allele and is a PKU carrier. The incidence of cystic fibrosis among Euro-Americans is about one in 2,500. This means that about one in twenty-five Euro-Americans is a carrier of the relevant allele.

Carrying one allele associated with a recessive condition is only a problem if one intends to have children with a partner who also has that particular allele. In that case, as we will see, each child will have one chance in four of inheriting copies of the allele from both parents and manifesting the condition, and one chance in two of being a symptom-free carrier. Carrying one copy of a recessive gene can actually be an advantage. For example, the allele implicated in sickle-cell anemia confers resistance to malaria on people who inherit only one copy of it. These people are said to have sickle-cell trait and exhibit no symptoms of sickle cell anemia. This resistance to malaria is believed to be the reason why the sickle-cell allele has become relatively prevalent among people native to equatorial Africa, where malaria has long been endemic. In the United States malaria is no longer a serious problem and the sickle-cell trait is irrelevant, except to couples who are both carriers.

We all carry alleles that would be disabling or lethal if we or our children had two copies of them instead of just one. Eugenic measures directed at people who manifest recessive conditions can only touch the tip of the iceberg. So, not only are there strong political and ethical arguments against instituting such measures but they cannot reduce the prevalence of such conditions in the population at large.

Although all this was known by 1908, before eugenic programs were instituted in the United States or in Germany, the old eugenic practices went on for several more decades, until changes in the political situation made them unacceptable. Yet the root concepts of eugenics survived the demise of its earlier forms. The idea of "race purity" may have died; the idea of building a strain of supermen may have died; but the idea that it is more beneficial for certain people to have children than others, and that a vast range of human problems can be cured once we learn how to manipulate our genes, remains very much with us.

Eugenics can take on many guises. Helen Rodriguez-Trias, currently (1993) president of the American Public Health Association, cites a 1972 survey of obstetricians which found that "although only 6 percent favored sterilization for their private patients, 14 percent favored it for their welfare patients. For welfare mothers who had borne illegitimate children, 97 percent . . . favored sterilization."[1] This is classic eugenic thinking, but eugenics can appear in much subtler ways. Any suggestion that society would be better off if certain kinds of people were not born puts us on a slippery slope.

Testing prospective parents to see if they are carriers of genetic "defects" leads to the labeling of large groups of people as "defective." Not only the people who manifest the condition but also the carriers are likely to be considered less than perfect. Such tests are generally considered to be altogether helpful because they increase people's choices, but it would be a mistake to ignore the ideology that almost inevitably accompanies their use.

In 1971, Bentley Glass, retiring as president of the American Association for the Advancement of Science, wrote:

> In a world where each pair must be limited, on the average, to two offspring and no more, the right that must become paramount is . . . the right of every child to be born with a sound physical and mental constitution, based on a sound genotype. No parents will in that future time have a right to burden society with a malformed or a mentally incompetent child.[2]

In a similar vein, the theologian Joseph Fletcher has written: "We ought to recognize that children are often abused preconceptively and prenatally—not only by their mothers drinking alcohol, smoking, and using drugs nonmedicinally but also by their *knowingly* passing on or risking passing on genetic diseases."[3] Notice that Fletcher absolves doctors of responsibility, singling out "nonmedicinal" drug use. This language of "rights" of the unborn implicitly translates into obligations of future parents, and especially future mothers.

This logic becomes explicit in the writings of Margery Shaw, an attorney and physician. Reviewing what she calls "prenatal torts" (a term I believe she invented), she argues as follows:

> Once a pregnant woman has abandoned her right to abort and has decided to carry her fetus to term, she incurs a "conditional prospec-

tive liability" for negligent acts toward her fetus if it should be born alive. These acts could be considered negligent fetal abuse resulting in an injured child. A decision to carry a genetically defective fetus to term would be an example. . . . Withholding of necessary prenatal care, improper nutrition, exposure to mutagens and teratogens, or even exposure to the mother's defective intrauterine environment caused by her genotype . . . could all result in an injured infant who might claim that his right to be born physically and mentally sound had been invaded.[4]

What is this "right to be born physically and mentally sound?" Who has such a right and who guarantees it? Shaw assumes not only that a fetus has rights (a hotly debated assumption) but that its rights are different, and indeed opposed to, those of the woman whose body keeps it alive and who will most likely be the person who cares for it once it is born. What is more, she places the burden of implementing these so-called rights of fetuses squarely on the shoulders of individual women.

Shaw does not suggest that women must have access to good nutrition, housing, education, and employment so that they are able to secure a fetus its "right" to proper nutrition and avoid its being exposed to mutagens and teratogens. She only urges that "courts and legislatures take all reasonable steps to insure that fetuses destined to be born alive are not handicapped mentally and physically by the negligent acts or omissions of others." Her language of "rights" does not advocate the kinds of improvements that would benefit women and children. It is the language of eugenics and social control.

Such control is advocated explicitly by John Robertson, a professor of law at the University of Texas (the same faculty at which Shaw teaches). His basic proposition is this:

The mother has, if she conceives and chooses not to abort, a legal and moral duty to bring the child into the world as healthy as is reasonably possible. She has a duty to avoid actions or omissions that will damage the fetus. . . . In terms of fetal rights, a fetus has no right to be conceived—or, once conceived, to be carried to viability. But once the mother decides not to terminate the pregnancy, the viable fetus acquires rights to have the mother conduct her life in ways that will not injure it. . . . The behavioral restrictions on pregnant women and the arguments for mandating fetal therapy and prenatal screening illustrate an important limit on a woman's freedom to control her body during pregnancy. She is free not to conceive, and free to abort after conception and before viability. But once she chooses to carry the child to term, she acquires obligations to assure its well-being. These obligations may require her to avoid work, recreation, and medical care choices that are hazardous to the fetus. They also obligate her to preserve her health for the fetus's sake or even allow established therapies to be performed on an affected fetus. Finally, they require that she undergo prenatal screening where there is reason to believe that this screening may identify congenital defects correctable with available therapies.[5]

Clearly the coercive mind-set underlying the old eugenics has not disappeared. Hardly anyone now uses old phrases like "racial deterioration," but casting the issue in individualistic terms and advocating for the rights of the unborn can be just as coercive. It converts the so-called choices of future parents, especially of mothers, into obligations to make a socially approved choice.

This is the spirit in which to understand the statement by Paul Ramsay, a theologian at Princeton University, that "the freedom of parenthood is . . . not a license to produce seriously defective individuals" or the recommendation by the Chicago Bar Association that Illinois require premarital tests for "diseases or abnormalities causing birth defects" before issuing marriage licenses.[6]

Genetic researchers often justify requests for funding by stressing the economic costs of caring for disabled babies. Such eugenic concerns frequently hover in the background of statements scientists, physicians, and genetic counselors make, even when they claim only to be interested in the individuals who manifest a genetic condition or who believe their offspring to be at risk for one.

PARENTING, DISABILITIES, AND SELECTIVE ABORTION

Relatively few diseases or disabilities are genetic, even fewer can be predicted, and most of the risks we and our families encounter are not biological at all. We expose ouselves to greater risks by living in cities and allowing our children to ride bicycles than by foregoing prenatal tests, unless we know that a specific condition runs in our family. Even then, in most instances tests can only reveal that there is likely to be a problem, not whether the child will be mildly or severely disabled.

Yet, among affluent women in the United States, predictive tests are becoming part of routine prenatal care. Such tests are a source of regular income for drug companies, hospitals, and private physicians. Also, doctors do not want to leave themselves open to charges of negligence, so when tests are available they recommend using them, even if there is no specific reason to expect problems.

The benefits of these tests for prospective parents are not so clear. Having children is always chancy and many things can go wrong that have nothing to do with genes. Prospective parents need information and resources in order to make realistic decisions about whether to bother with prenatal testing, or whether to carry a pregnancy to term even though tests show that their future child will have a particular condition. When counselors inform future parents that the incidence of a given inherited condition is one in a hundred, they often do not feel the need to point out that the incidence is *only* one in a hundred, and that chances are ninety-nine to one that the child will *not* manifest that condition.

For most of us, statistics have little meaning when it comes to our health or the health of our children. We hope for the best and fear the worst, whatever the numbers. As my mother, a physician, used to say, "If it happens to me, it's a hundred percent."

This is not just foolishness. When we are talking about such personally serious issues, knowing the probabilities is of little use. None of us can be guaranteed that things will turn out the way we want and, if they don't, it is no comfort to know that the percentages were in our favor. Good genetic counseling must give people information about the medical, social, and educational resources they may need, and how accessible these resources are. Perhaps the most important thing a counselor can do is to tell people how to find parents of children who have the disability their child may have, so that they can establish peer support as early as possible.

The mind-set behind genetic testing rests on societal views of disabilities that should not go unchallenged. Many of the difficulties people with disabilities experience are not intrinsic to their physical or mental state, but result from societal obstacles that could be removed by appropriate social or economic measures. Just as women (whatever their race) and men from minority groups have been excluded from many occupations and professions, irrespective of their individual ability to perform the tasks involved, people with inherited or acquired disabilities have been denied entry into many schools and jobs.

The presumption has been that people with mental or physical disabilities cannot perform adequately. Yet the effects of various disabilities differ in their extent, and even people who have the same disability may experience it quite differently. Stephen Hawking, the physicist and author of the best-selling *A Brief History of Time*, who is severely disabled by amyotrophic lateral sclerosis (ALS), or Lou Gehrig's disease, has said that, owing to a career involving primarily mental work, and unfailing support from family and colleagues, "My disability hasn't been a serious handicap."[7]

Being blind is not at all like being deaf, or like having a painful disease or a mobility problem, and not every blind (or deaf) person has the same capabilities and constraints. Many genetic conditions vary in their severity and often their symptoms can be alleviated, at least to some extent, by conventional medical therapies. A prenatal diagnosis of sickle-cell anemia or cystic fibrosis does not predict at what age the condition will become manifest, how disabling it will be, or to what extent it will shorten the life of the affected individual. As more effective therapies have become available, the quality of life for people with these conditions and for their families has improved markedly.

Down syndrome, which is caused by the presence of an extra chromosome, was thought until recently to preclude a person's attending school, holding a job, or functioning in society. Yet, in the past ten years, determined parents and educators have developed educational methods that have enabled many people with Down syndrome to read, write, and acquire a range of skills. They can hold jobs, provided that people hire them, and can establish strong and lasting relationships. Even with predictive tests, prospective parents of children with Down syndrome can know little about what to expect—which is not so different from the situation of any future parent.

All people with noticeable disabilities share the oppression and discrimination directed at them by our culture, because in this society we learn since childhood that to be dependent or sick is to be less than wholly human. This negative perception is often based on ignorance. A recent study of pregnant women, asking whether they wanted to be screened for cystic fibrosis and whether they would continue a pregnancy if their fetus were predicted to have the condition, found that the women who knew more about cystic fibrosis were less frightened about it than the women who were unfamiliar with this condition.[8]

The eugenic ideology ignores the shadings and interactions that make certain human characteristics into disabilities. Marsha Saxton, a disability rights activist who was born with spina bifida, a neurological condition for which pregnant women are routinely screened, sums up the assumptions underlying much prenatal testing as follows:

> (1) that having a disabled child is a wholly undesirable thing, (2) that the quality of life for people with disabilities is less than that for others, and (3) that we have the means ethically to decide whether some people are better off never being born.[9]

We need to recognize these assumptions when we meet with them, and be ready to challenge them.

Despite the eugenic implications of prenatal testing, if tests are available then women must have the option to take or refuse them. However, what does such an option mean in a society like ours, which offers far too little medical and social support to parents of children with disabilities? To make real choices possible requires more than merely to permit them.

I unequivocally support a woman's right to terminate a pregnancy, whatever her reasons. Despite the problems involved in prenatal testing, a prospective mother who decides to be tested must be able to act on the results in whatever ways fit her circumstances. That means that she must have the right either to abort or to continue the pregnancy, without external pressure. Unfortunately, such pressures are everywhere. Many women who wish to have an abortion cannot get one, for financial or other reasons. Conversely, many women who can pay for tests are considered irrational if they decide not to have an abortion when the fetus they are carrying has been diagnosed as having a disability.

I want to stress that I see a marked difference between a woman having an abortion because she does not want a child and having an abortion, though she wants a child, because she does not want this one. Confronting parents with such a choice is uncomfortably like asking them to play God. After all, with or without technology, every pregnancy is wreathed in uncertainty.

Most prenatal tests offer little precise information. They can suggest problems, but cannot say how significant these problems may be. Genetic predictions, like all medical tests, involve setting arbitrary norms. People, or fetuses, who fall outside them are by definition "abnormal," irrespective of whether they exhibit noticeable symptoms or whether these symptoms are particularly debilitating. Women are aborting fetuses because physicians have diagnosed a chromosomal irregularity, even though no one can say whether this irregularity would have noticeable effects. Those who refuse to accept such definitions often face reactions ranging from disbelief to hostility.

Recently, Jane Norris, a radio talk-show host in Los Angeles, conducted a call-in program in which she expressed outrage at Bree Walker Lampley, a well-known CBS television news anchor in Los Angeles, and encouraged her callers to do the same. Walker Lampley and her young daughter have a heritable condition called *ectrodactyly*, and she was expecting a second child. People with ectrodactyly often cannot freely move their fingers and toes, because some of the bones in their hands and feet are fused. Though this condition clearly has not prevented Walker Lampley from pursuing a productive public career, Norris insisted on referring to the condition as a "deforming disease." By bringing such a "deformed" human being into the world, she asked repeatedly, wasn't Walker Lampley being unfair to society and the future child?[10]

Because Walker Lampley is a public personality, this story received a lot of media attention. However, people are encountering similar prejudices all the time. Paul Billings, chief of the Division of Genetic Medicine at Pacific Palisades Hospital in San Francisco, and a group of researchers at Harvard Medical School have been conducting a study of genetic discrimination in the United States.[11] In one instance they cite, a woman who had a child with cystic fibrosis discovered on the basis of prenatal tests that the fetus she was carrying would also develop cystic fibrosis. When she decided to continue the pregnancy, the health maintenance organization (HMO) where her family was receiving its medical care informed her that, while it would pay for an abortion, it would not cover her further prenatal care or the medical care of the child she would bear, as the child now had a "preexisting condition" that was not eligible for insurance. The HMO changed its stance only after vigorous protests.

The idea that people "like that" should not be born is based on stereotypes of who people "like that" are. People have speculated that Abraham Lincoln had Marfan syndrome, a dominant condition for which scientists are now developing a predictive test.[12] Would the world be better off if Lincoln had not been born?

Many policies, laws, and proposed laws that restrict (or would restrict) abortion in the United States and abroad do permit eugenic abortions performed on grounds of fetal disabilities. Such policies, based on the notion that people "like that" should not be born, illustrate the depth of the prejudice against people with disabilities. And yet, all of us can expect to experience disabilities—if not now, then some time before we die, if not our own, then those of someone close to us. If only for our own good, we must dispel the dread of disability that motivates such pervasive prejudices, and so limits the lives of many people.

As I have said before, many genetic conditions are extremely variable. The most common condition that is inherited as a Mendelian dominant is Huntington disease, a slowly progressing, degenerative condition of the nervous system that can involve disorientation, mental deterioration, and, eventually, death. Most people with the Huntington allele do not begin to develop symptoms until mid-life or even old age, though a few exhibit them as children.

Although scientists have not yet isolated the Huntington allele itself, they have identified a set of *markers*, sections of DNA that are near the gene, and can test people for the presence of those markers. Since scientists cannot look directly at the gene, they cannot simply give someone a test and discover whether that person has the mutation. They first need to compare the *marker DNA* of as many of that person's family members as possible in order to identify features that are common to all relatives who have the condition and that are not shared by any who do not have it and are old enough that they probably will not get it. Therefore, someone who wants to be tested for the presence of the Huntington mutation must have the cooperation of many family members, some of whom have the condition and some of whom do not.

It is becoming clear that by no means everyone who may have inherited the Huntington allele wants to be tested so as to know in advance whether or not he or she will develop the condition. Woody Guthrie, who died of Huntington disease, lived a productive life and left a legacy of over a thousand wonderful songs. His son Arlo has said in interviews that he does not want to be tested, though he has a 50 percent chance of having inherited the mutation implicated in this condition. He sees nothing irresponsible in having had three children, each of whom will have a 50 percent chance of developing the condition if he develops it. As all of us die sooner or later, he feels that the point is to contribute to society while we live, rather than worry about when death will come, or from what cause.

It is interesting to see the way genetic testing and selective abortions are discussed outside the United States. A particularly interesting example is Germany, which still lives in the shadow of the Nazi eugenics programs. An article in the British science weekly *Nature* reports that the German Parliament has limited the use of genetic screening tests. The same article states that James Watson, of the U.S. genome project, called this action "backward" and questioned why a developed nation on such occasions should persist in "invoking the name of Hitler."[13]

Watson's criticism implies that German scientists under Hitler were somehow different from scientists in other times and places. Unfortunately, as many Germans remember, the Nazi programs of eugenic "selection and eradication" were designed and put in place by respected and respectable academics, jurists, and heads of hospitals and scientific institutes. These people would be hard to distinguish from their present-day successors. They were just operating in a different political climate.[14]

In this country, too, we must be vigilant and deliberate about what lines to draw and about who gets to draw them. Such decisions must not be left to the technical experts—the medical geneticists and molecular biologists—who have a professional and financial interest in the outcome. They need to be integrated into the political process and made in the context of decisions about access and financing of medical and social programs.

GENETIC SCREENING

So far, I have been talking mostly about genetic tests, which prospective parents can use to find out whether their child is likely to have a specific inherited condition. Genetic screening is different because it involves testing populations rather than individuals who may be concerned about their or their children's health.

While it is impossible to screen everyone for every gene that might be involved with a disease or disability, many people advocate routinely testing groups among whom a specific inherited condition is unusually prevalent. For example, in the early 1970s scientists developed a relatively simple test to detect carriers of the Tay-Sachs allele. The incidence of Tay-Sachs disease is one per 100,000 in the entire U.S. population, but it occurs at a rate of about one per 3,600 among Ashkenazi Jews (Jews of east European origin). Therefore, campaigns have been launched among Ashkenazi Jews in the United States, Canada, Great Britain, Israel, and South Africa to detect Tay-Sachs carriers and notify couples in which both partners are carriers. This enables such couples to monitor their pregnancies and detect a fetus that might have the condition. On the whole, these programs have met with support from Jewish communities.

Because Tay-Sachs screening programs have been generally welcomed and successful, they are often held up as an example of the benefits of genetic screening. However, Tay-Sachs is rather a special case. To date there is no therapy or cure for this condition and it is invariably fatal in the first years of life. The extreme severity of this condition and the fact that the at-risk population is relatively cohesive, educated, and economically independent, gives the Tay-Sachs programs a major advantage.

This is in contrast to sickle-cell screening, which got off to a bad start in the early 1970s. About one in five hundred African Americans has two copies of the sickle-cell allele and is therefore likely to develop symptoms of sickle-cell anemia. Many more, about one in ten, are carriers for sickle-cell trait, which can be detected by blood tests that have been available since the 1960s.

In the early 1970s, when civil rights activists highlighted the disparities in health status and mortality rates between Euro- and Afro-Americans (disparities which are almost all societally caused, and which have increased since that time), President Nixon responded by directing special attention to sickle-cell anemia. Nixon had already declared a "war on cancer." Why not show concern for civil rights by also declaring war on a "black disease"? (Of course, neither his nor subsequent Republican administrations have instituted the economic and social measures that would be required to improve the health of African Americans across the board.)

In 1972, the U.S. Congress passed the Sickle-Cell Anemia Control Act and several states promptly instituted screening programs. Sickle-cell anemia itself is usually diagnosed because it produces symptoms, so the screening programs were intended to detect carriers of the sickle-cell trait. There was no provision for counseling, and often the difference between the disease and the carrier state was not made clear to the people who were being tested. As a result, people who were carriers became needlessly concerned about their health. In some places, sickle-cell tests were made compulsory on entering school or before couples could get a marriage license, and some programs required people to pay for their own tests.

In the early and mid-1970s "almost all of the major airlines grounded or fired their employees with sickle-cell trait."[15] The U.S. Air Force Academy also instituted a policy to exclude sickle-cell carriers, until one trainee filed a lawsuit in 1979. All this was done on the unproved assumption that such carriers were less able than other people to withstand the stress of lowered oxygen levels at high altitudes. Some African-American sickle-cell carriers reported being denied employment or insurance, or having their insurance premiums raised.

Because of vocal criticism, including accusations of genocide, large-scale screening for sickle-cell trait stopped by the mid-1970s. However, the test used for this screening did not detect the sickle-cell allele, only the presence of sickle-cell hemoglobin in the blood. In the early 1980s, a reliable test became available that could detect the allele itself. This opens the possibility of predicting the trait or the anemia in a fetus. Therefore, there have been new calls to screen African Americans for the allele, both from people within the African-American community and from outsiders. This would give parents who are both carriers the opportunity to have each fetus tested early enough to be able to decide whether to continue the pregnancy.

In 1976, Congress passed the National Genetic Diseases Act, which provides for research, screening, counseling, and professional education for people involved with Tay-Sachs disease, cystic fibrosis, Huntington disease, and a number of other conditions in which gene mutations are implicated. From time to time, organizations of people with one or another inherited disability or disease lobby to have their condition added to the list.

As the failure of the sickle-cell screening programs shows, any program that raises doubts about people's genes must be handled very delicately. People must be educated to understand that carrier status is of no interest, except when two carriers for the same condition decide to have children together. Even with Tay-Sachs programs, which have been particularly successful, it is important to be sure that screening is preceded by adequate counseling and that people who do not wish to be tested or who decide to continue a pregnancy despite a prediction of the disease are not made to feel irresponsible, stupid, or crazy. This sort of care is hard to provide even when the population to be screened is relatively small and well defined. To devise responsible screening programs for the majority of people in a large country is virtually impossible.

In the United States, cystic fibrosis is the most common condition with a predictable pattern of inheritance, affecting about one in 2,500 Euro-Americans. Over forty different alleles have so far been associated with cystic fibrosis, and recently a diagnostic test has been developed for the most common one.[16] Since this allele accounts for about 75 percent of the incidence of the condition, there have been calls for mass screening programs to detect its carriers.

In 1990, two physicians, Benjamin S. Wilfond and Norman Frost, published an article in the *Journal of the American Medical Association* in which they evaluated the advisability of such a massive screening program.[17] They estimate that even for a relatively prevalent condition like cystic fibrosis, population-wide screening to detect unsuspecting carriers would cost more than one million dollars for each child who might otherwise develop the condition. A genetic screening program of this magnitude would vastly exceed the resources of available genetic counselors and medical geneticists. Since, according to the Hardy-Weinberg theorem, about one in twenty-five Euro-Americans is a cystic fibrosis carrier, such a program would inevitably spread confusion and anxiety among millions of healthy people who have no reason to be concerned about their carrier status.

Wilfond and Frost recommend against a cystic fibrosis screening program. They emphasize that it will be important for policymakers to resist pressure from the biotechnology industry, which could reap enormous profits from a program of this magnitude, as well as from individuals who may become unduly optimistic about the benefits of knowing whether they are cystic fibrosis carriers. They urge that the safety and effectiveness of such a screening program first be explored in small pilot studies and warn that "the ability to accurately determine the carrier status of an individual should not inherently imply that population screening ought to be adopted as public policy or even be left to free market forces."[18] Despite this and similar warnings, Colorado now requires that all newborns be routinely screened for the gene mutation implicated in cystic fibrosis.[19] In his book, *Backdoor to Eugenics*, the sociologist Troy Duster points out that the very term "screening" implies the existence of something bad which one needs to guard against.[20] It is therefore virtually impossible to screen populations without stigmatizing some people as "defective" or "abnormal."

Molecular biologists argue that, because the genetic tests they are developing will show that all of us are flawed in one way or another, these tests will bring an end to genetic discrimination. This claim is disingenuous. In an unequal society like ours, different kinds of people experience disabilities and discrimination differently, depending on how they are labeled and how they are perceived.

In the 1930s, during the heyday of the eugenics movement, the British geneticist J. B. S. Haldane pointed out that although hemophilia was known to be prevalent in the royal houses of Europe (apparently introduced into Great Britain and thence into continental Europe by none other than Queen Victoria), no one was suggesting that members of the royal families be sterilized.[21] Similarly, it is not surprising that, in the United States, African Americans have been the prime group to experience genetic discrimination. Like other forms of discrimination, genetic discrimination will be felt most by people who are already stigmatized in other ways. People with access to power and resources are more likely to be shielded.

FALLACIES OF GENETIC PREDICTION

Genetic predictions, whether they involve testing or screening, are based on the assumption that there is a relatively straightforward relationship between genes and traits. However, genetic conditions involve a largely unpredictable interplay of many factors and processes. To quote the authors of the popular genetics textbook *An Introduction to Genetic Analysis*:

> A gene does *not* determine a phenotype [noticeable trait] by acting alone; it does so only in conjunction with other genes and with the environment. Although geneticists do routinely ascribe a particular phenotype to an allele of a gene they have identified, we must remember that this is merely a convenient kind of jargon designed to facilitate genetic analysis. This jargon arises from the ability of ge-

36

neticists to isolate individual components of a biological process and to study them as part of genetic dissection. Although this logical isolation is an essential aspect of genetics, . . . a gene cannot act by itself.[22]

Even genes that are implicated in conditions whose inheritance follows a regular and predictable pattern are proving to be far from simple to define and localize. For example, the gene associated with Huntington disease, which is thought to lie on chromosome 4, has so far resisted precise localization or analysis. In fact, some scientists are beginning to wonder whether DNA in more than one region of this chromosome may be involved.

Similarly, identification of "the cystic fibrosis gene" and its localization on chromosome 7 is running into unanticipated complications. As I have mentioned, many different mutations appear to be associated with this condition in different individuals. In fact, cystic fibrosis probably is not a single entity, but a group of related conditions with somewhat different manifestations that result from different mutations in the DNA sequence.[23]

To provide meaningful genetic information, scientists may sometimes need to work out the pattern of mutations separately for different families or even for different individuals who manifest the "same" disease. This would make predictions impossible. Most inherited conditions exhibit a variety of symptoms and patterns of development, and may turn out to be families of related conditions rather than unique entities. This is especially likely when it comes to highly variable conditions of late onset. Researchers have made repeated claims (later contradicted) to have identified "the Alzheimer gene." Lately, two different groups of scientists have identified two different pieces of DNA, one on chromosome 19 and the other on chromosome 21, each group claiming that their "Alzheimer gene" is the real one.[24]

The situation becomes even more complicated when scientists try to predict conditions that are said to involve inherited "tendencies." It is becoming conventional to argue that whatever chronic condition we die of—cancer, coronary heart disease, diabetes, stroke, or alcohol- or drug-induced liver disease—is likely to be "determined" by tendencies we inherit in our genes. So, the search is on for genes "for" cancer, cardiovascular disease, and even behavioral conditions such as alcoholism.

From a therapeutic perspective, it makes little sense to try to sort out the genes involved with complex genetic conditions, even if DNA is involved at some level. Yet, the faith that genes for all sorts of troublesome conditions can be identified and isolated, coupled with the hope that this will lead to profitable diagnostic tests, is likely to continue to fuel the search for relevant bits of DNA. Not only will this not cure or prevent the conditions, it will create a new group of stigmatized people, the "asymptomatic" or "healthy ill" who, though they have no symptoms, are considered likely to have a particular disability at some point in the future.

. . . .

FOUR

A BRIEF LOOK AT GENETICS

HEREDITY AND GENES

Our ideas about heredity and about the ways genes function have been influenced by the way we look at one another. We tend to note differences among ourselves more than similarities. If asked to compare Scandinavians and West Africans, we say Scandinavians are tall, fair-skinned, blond, and blue-eyed, while West Africans are dark-skinned, with curly hair, and dark eyes. We do not mention that both have two arms and two legs attached to a torso, with a head that contains a mouth, a nose, two ears, and two eyes. We know what human beings look like. So, in comparing groups of humans, we describe the ways in which they differ.

This method of description by difference only works if one already knows the similarities. To some creature who had never seen a person, one would have to describe Africans and Scandinavians in almost the same words. To start the description by saying one kind of person is dark-skinned while another kind is light-skinned, without first explaining what a "person" is, would convey little information.

And yet, that is the basis of genetics. From its earliest days, this science has focused on how organisms differ, not on their similarities. If mutations didn't happen, if traits didn't pop up unexpectedly, scientists might never have asked the kinds of questions that led them to postulate the existence of genes, or to look for them in the chemical substances that make up cells.

As the British geneticist J. B. S. Haldane wrote half a century ago:

> Genetics is the branch of biology which is concerned with innate differences between similar organisms. . . . Like so many other branches of science, genetics has achieved its successes by limiting its

scope. Given a black and a white rabbit, the geneticist asks how and
why they differ, not how and why they resemble one another.[1]

Recently, molecular biologists have also begun to study likeness—
how rabbits succeed in always producing rabbits. Historically, however,
geneticists have asked only why two black rabbits sometimes can produce a white rabbit, not why they produce rabbits instead of guinea pigs.

THE BEGINNINGS: GREGOR MENDEL, "TRAITS" AND "FACTORS"

Looking at the history of genetics will help us get a sense of where the
facts and concepts of contemporary genetics come from, and see how the
concept of the gene arose. The usual story begins with Gregor Mendel,
a Czech monk, and his experiments breeding garden peas in the monas-
tery garden at Brno. In 1865, Mendel published his results in a single,
classic paper, in which he laid out what have come to be known as
Mendel's laws.[2] Mendel's paper excited little interest at the time, but was
rediscovered around 1900 and became the basis for modern genetics.

Mendel did not discover the gene, as he is sometimes said to have
done. As he tended his peas, he was not even interested in what went on
inside them. He did not ask why or how they produced others like themselves. He did not describe inheritance; he described change. He tried to
understand patterns of difference between parents and offspring, and did
not concern himself with causes.

Mendel concentrated on visible traits that his plants exhibited—the
colors of their flowers and the shapes and textures of their seeds. Only
once did he refer to hypothetical "factors" inside the plants that might
correspond to these traits. His concern was with what the Danish bota-
nist Wilhelm Johannsen later called the *phenotype*, the organisms' outward characteristics, not their *genotype*, the genetic make-up underlying
those characteristics that they can pass on to their offspring.

Mendel was working before scientists were thinking about chromosomes or genes. He was interested in the breeding patterns of hybrids,
and his laws are mathematical descriptions of the extent to which certain
traits occur in successive generations. He noted that pea plants with red
flowers are of two kinds. When he crossed two plants of the first kind,
they always produced plants with red flowers: they *bred true*. However,
when he crossed the other kind of red-flowered plants, three out of four
of the offspring were red-flowered, but one quarter had white flowers.

In contrast, the white-flowered plants always bred true, only producing other white-flowered plants.

To explain this and similar phenomena, Mendel symbolized the true-breeding plants with a capital **A** for red-flowered plants and a lower-case **a** for white-flowered plants. The red-flowered plants that can give rise to plants with white flowers he assumed were hybrids and called **Aa**. He explained the phenomena he observed by saying that when two **A**-plants are crossed, all their offspring will be **A** (red flowers) and, similarly, when two **a**-plants are crossed, their offspring will all be **a** (white flowers). But when the red hybrids (**Aa**) are crossed with each other, one in four of their offspring will be **A** (true-breeding red), two in four (one-half) will be **Aa** (red hybrids), and one in four will be **a** (true-breeding white). The three-to-one ratio of red to white-flowered plants, which he had observed, masks the existence of three different kinds of plants, **A**, **Aa**, and **a**, in the ratio $1:2:1$. Three quarters of the crosses produce either **A** or **Aa** plants, which look the same. One can only tell them apart by breeding them. (Modern geneticists, to make this clearer, call the **A** plants **AA** and the **a** plants **aa**.)

Since the rediscovery of Mendel's laws around 1900, traits that are passed on in this straightforward and predictable manner are called *simple Mendelian traits*. By now many such traits have been described, but it is important to realize that they constitute only a small fraction of observed traits.

In the language of modern genetics, we say that Mendelian patterns of inheritance are mediated by different *alleles,* or variants, of the same gene. In the example of Mendel's red and white flowers, **A** and **a** are alleles of the gene that mediates flower color. Allele **A** mediates the plant's ability to synthesize a red pigment, while allele **a** leaves the flowers unpigmented. Plants that have allele **A**, whether they are **A** or **Aa**, will produce red flowers.

Mendel called traits that behave like **A** dominant, because they mask the alternative **a**. Traits like **a**, which only appear when not masked (as in an **aa** plant), he called *recessive*. Later scientists, who focused on genes rather than on traits, have applied the same nomenclature to genes, so that we now speak of dominant and recessive alleles of a given gene.

Although most traits do not follow the patterns of inheritance described by Mendel, scientists, believing that many of these are nonetheless mediated by genes, have devised more complicated genetic explanations. One is that the trait is *polygenic*, that is, "controlled" by so many genes that the Mendelian pattern gets washed out. This assumption, that many genes are involved where one will not explain the observations, was criticized in the early days by the U.S. geneticist, Thomas Hunt Morgan. As he wrote in a paper for the American Breeders Association:

> In the modern interpretation of Mendelism, facts are being transformed into factors at a rapid rate. If one factor will not explain the facts, then two are invoked; if two prove insufficient, three will sometimes work out.[3]

Morgan and his colleagues also warned that "a single factor may have several effects, and that a single character may depend on many factors."[4] It will be well to remember this warning when we come to look at conditions that are said to follow complex patterns of inheritance involving many genes.

FROM MENDEL TO THE DOUBLE HELIX

Mendel's paper excited little attention and was soon forgotten. By the turn of the century, when it was rediscovered, the mainstream of scientific interest among biologists had begun to shift from describing organisms and their parts (taxonomy, anatomy) to trying to understand how organisms function (physiology, embryology). The German scientist August Weismann postulated the existence of what he called the *germ plasm*, a material substance in eggs and sperm that in some way carried heritable traits from parents to their offspring.

By the latter part of the nineteenth century, scientists had learned how to stain cells and so make visible certain structures inside them, which they called *chromosomes*, and to watch the transformations these chromosomes undergo during cell division. They saw that, just before a cell divides, the chromosomes line up and duplicate. The two sets of chromosomes move apart as the single cell divides, so that each of the new cells has its own complete set of chromosomes.

Weismann suggested that these chromosomes are the bearers of heredity. In 1909, Johannsen coined the word *gene* to denote hypothetical particles that are carried on chromosomes and mediate inheritance. Though somewhat more specific than Mendel's "factors," the new word still represented an idea more than an object. As Morgan put it in 1926: "In the same sense in which the chemist posits invisible atoms and the physicist electrons, the student of heredity appeals to invisible elements called genes."[5]

Chromosomes contain both proteins and DNA (*deoxyribonucleic acid*). For many decades an active and often heated debate raged over which of these was the bearer of heredity. Proteins were favored by many scientists, because their structure is more varied and complex than that of DNA and they therefore looked to be the more likely carriers of the wide variety of traits that get transmitted to successive generations.

Experiments done in the mid-1940s and early 1950s settled the matter by showing that when DNA obtained from bacteria or viruses is introduced into bacterial cells it can transform these cells so that they exhibit traits characteristic of the donor-cells from which the DNA was obtained. By 1953, when James D. Watson and Francis Crick proposed their model of the structure of DNA, the "double helix," biologists were agreed that DNA is the material that mediates heredity.

This did not mean that they had "found" the hypothetical genes. No matter how one may look at DNA, there are no discrete little balls that carry hereditary traits. Rather, specific traits appear to be mediated by sections of DNA, which is a long, threadlike molecule. It is an easy shorthand to call these segments "genes," as they do seem to serve the functions geneticists assigned to those theoretical particles.

In a sense, the "gene" no longer has a physical meaning for molecular biologists. The material reality is DNA. But since genes remain very much a part of the science of genetics, as well as of the culture at large, experiments with DNA get communicated in the language of genes. Sometimes this raises problems, when what scientists observe with DNA does not readily fit into a model of discrete particles. Still, molecular biologists often feel that it is easier to explain their work if they perpetuate the model of little balls on a string, passed on from parents to children.

We might be better off if people understood what scientists do and do not know about the role DNA plays in the metabolism and growth of organisms, without enveloping this knowledge in clouds of gene-talk. However, DNA and genes are so intertwined in the scientific and popular imagination that it is not always easy to be sure which we are talking about.

GENES AND PROTEINS

All we know for sure about the function of DNA is that segments of it can specify the linear sequence of the *amino acids* that are the basic components of proteins, and that other parts of DNA help specify when and how fast the proteins will be synthesized. The reason this is important is that proteins are involved in everything that happens in living organisms.

Cells contain a variety of proteins at every level of their structure.

There are proteins in the membranes that envelop the cells, in the protoplasm that surrounds the nucleus, in the nucleus itself, and in the chromosomes. The body fluids that bathe our cells and tissues contain still other proteins.

All these proteins are different and have different functions. Some provide structural support, some help interpret the message in the DNA, some act as channels to allow substances (sometimes other proteins) to get in and out of cells. Hair, nails, and feathers are made of *keratin,* a protein. Egg white consists mainly of *albumin,* a protein. Muscles contract or relax through motions of the proteins in their cells. Hemoglobin, the red pigment in blood that transports oxygen and carbon dioxide, is a protein, as are the antibodies in our immune system and many of our hormones.

When Frederick Engels wrote in 1878 that "life is the mode of existence of albuminous substances [proteins],"[6] he was taking a rather reductionist view, but it is true that proteins are involved in everything that happens within living organisms. Scientists believe that proteins and DNA were around before there was life, and that it was reactions among them that produced the first life forms.

When people talk about genes mediating traits such as eye color or inherited medical conditions, they mean that those traits arise from the activities of proteins whose composition is specified by those particular genes. However, even the simplest traits involve not only a variety of proteins, but also other factors, both within and outside the organism. It is an oversimplification to say that any gene is "the gene for" a trait. Each gene simply specifies one of the proteins involved in the process.

How Chromosomes and Genes Are Duplicated

Human beings receive a set of twenty-three chromosomes from each parent. These chromosomes differ in size and shape and are identified by numbers from one to twenty-three. Identical sets of these forty-six chromosomes are located inside the nucleus of every cell in our bodies, and carry the functional units called genes. That means that we inherit two complete sets of genes.

When the nucleus of my mother's egg fused with the nucleus of my father's sperm, the nucleus of this new cell, which would eventually become Ruth, contained forty-six chromosomes in twenty-three pairs. These have been passed on with each cell division. Therefore, every cell nucleus in my body contains not just the same number, but the same kinds of chromosomes and genes as the nucleus of that original fertilized egg.

Actually, things do not always go quite that smoothly. Sometimes when the chromosomes are duplicated, a gene is copied slightly wrong. Such events are called mutations. I live in a world full of radiation and chemicals that may produce changes—mutations—in my DNA, which will then be duplicated at each subsequent cell division. Because of these mutations, here and there in my body there may be a nucleus that's a bit different from the nucleus of that original future-Ruth, but by and large all my nuclei are alike.

Of course, that does not mean that all my cells are alike. I have skin cells, blood cells, muscle cells, liver cells, and so on. The fascinating thing is that, although these different kinds of cells look different and perform different functions, they all have the same chromosomes and DNA in their nuclei. This is possible because, in any one cell, not all the genes are called upon to participate at any one time, or perhaps ever.

Given that my father and mother were different people, there must have been differences in the DNA they contributed to the nucleus of future-Ruth. To put it another way, they contributed the same kinds of genes, but sometimes different alleles, just as Mendel's pea plants all had the gene associated with flower color, but with different alleles for the different colors.

For any gene I have inherited, I may have gotten the same allele from each of my parents, or two different ones. If both gave me the same allele, I am said to be *homozygous* for that gene (homo is Greek for "same"). If they gave me two different alleles, I am said to be *heterozygous* for that gene (hetero is Greek for "different"). So, I have a patchwork of homo- and heterozygous pairs of genes in the nuclei of all my cells, and this patchwork is the same in every cell.

X AND Y: THE SEX CHROMOSOMES

Of the twenty-three pairs of chromosomes in our cells, twenty-two look the same in all of us. These are called *autosomes*. The twenty-third pair are called the *sex chromosomes* because they come in two shapes, which are different in women and men. In women, the two sex chromosomes have the same shape and are called X-chromosomes. Men have one X-chromosome and one Y-chromosome. The Y-chromosome is much smaller than the X-chromosome and contains many fewer genes. As with all my other pairs of chromosomes, I inherited one X-chromosome from my mother and the other from my father, but if I were a man I would have inherited an X-chromosome from my mother and a Y-chromosome from my father. This is because all eggs contain twenty-two autosomes and an X-chromosome, but sperm contains twenty-two autosomes and either an X- or a Y-chromosome.

Though we call them sex chromosomes, the X- and Y-chromosomes are not involved simply with sex-differentiation. Like all chromosomes, they carry a variety of genes. While at present the human Y-chromosome is not thought to carry many genes, the X-chromosome carries genes associated with numerous traits, such as color vision, blood clotting, and baldness.

CHROMOSOMES AND CELL DIVISION

Although there are forty-six chromosomes in each of our cells, in twenty-three matching pairs, we pass along only one complete set of twenty-three chromosomes to each of our children. The process by which this halving happens is called *meiosis*, and occurs whenever an egg or sperm cell is formed. As you will see, it has important consequences for inheritance between generations and for variations among the different children of any one pair of parents.

During an ordinary cell division (*mitosis*), the membrane that surrounds the nucleus disintegrates and the pairs of chromosomes line up along the mid-line of the cell. Each chromosome is duplicated and, as the cell constricts at its mid-line, a complete set of chromosomes migrates into each half. We end up with two *daughter cells*, each with its own newly formed nucleus.

As the chromosomes are duplicated, the DNA inside them must be duplicated as well. This replication occurs as follows: the two strands of the double helix in each chromosome unwind and each strand (let us call them **A** and **B**) becomes the template for the synthesis of its partner. That is, the cell uses strand **A** as a template to make a new complementary strand, identical to **B**, while the old strand **B** serves as a template for a new, complementary strand **A**. The old **A** strand then winds together with its new **B** partner, and the old **B** strand with its new **A** partner, creating two identical double helixes where before there was one. When the process of cell division is finished, each daughter cell has an identical set of daughter chromosomes, each containing identical double helixes of DNA. Thus, in mitosis, the entire complement of twenty-three chromosome pairs, with their genes, gets passed from each cell generation to the next.

Meiosis, from a Greek word meaning "to diminish," occurs only in the formation of the *gametes*, the eggs and sperm. In meiosis, the twenty-three pairs of chromosomes in the parent cell are replicated as before, but then the parent cell divides not just once, but twice. This is called a *reduction division*, because each of the four daughter cells receives only one set of twenty-three chromosomes rather than a pair.

In the generation of sperm, all four of these daughter cells become functional sperm. During egg formation, however, three of the four nuclei that are formed during meiosis are expelled as *polar bodies* and decay; only one is incorporated into an egg. Which of the four ends up being an egg nucleus is a matter of chance.

A point to bear in mind is that the chromosomes we inherit from each parent do not travel in sets. Therefore, during meiosis, when the forty-six chromosomes in a cell are replicated and divided into four sets of twenty-three, each set contains a mixture of paternal and maternal chromosomes. Each of my eggs contains a complete set of all the chromosomes from one through twenty-three, but it is a matter of chance whether any given chromosome came from my father or from my mother.

Actually, the story is even more chancy than that, because at the point at which the chromosome pairs pull apart and move into the daughter cells, they sometimes exchange comparable segments of DNA with each other. So, even the DNA of any given chromosome in my eggs can consist of a blend of genes I inherited from my two parents.

The fact that the twenty-three chromosomes are independent of one another when they sort themselves out during a reduction division, plus the fact that portions of them may have been exchanged between the two partners, makes it highly improbable that any two of my eggs will contain an identical set of chromosomes or genes. All the genes will have come from one or another of my parents (except for chance mutations), but the mix of my parents' genes within each of my eggs is a matter of chance.

FROM DNA TO RNA TO PROTEINS

We can picture the two strands in DNA as two ribbons consisting of alternating sugar and phosphate molecules: -sugar-phosphate-sugar-phosphate-. These ribbons are wound into a double helix and connected at regular intervals by horizontal rungs, each of which is formed by two *bases*, one attached to each ribbon. There are four kinds of bases in DNA: *adenine* (**A**), *thymine* (**T**), *cytosine* (**C**), and *guanine* (**G**). On any one strand of the double helix the four kinds of bases can occur in any sequence, but their shapes are such that, in order to fit together into rungs, every **A** on one strand must lie opposite a **T** on the other strand, and every **C** must lie opposite a **G**.

This geometric requirement is what always makes the two strands of the double helix complementary. Each single strand becomes the template for the other strand's synthesis, and the linear sequence of bases is copied at each cell division. If the sequence in a region of one strand is -A-C-C-A-T-G-, it automatically means that its complementary strand must have the sequence -T-G-G-T-A-C- at the corresponding region. The double helix and its replication are illustrated schematically in figure 1, but we must understand that, by focusing on DNA, this figure leaves out everything that makes the replication process possible—enzymes, other molecules, and a lot of interrelated metabolic activities.

The sequence of bases in a particular functional unit of DNA (a gene) is translated into a sequence of amino acids that makes up a protein. Each amino acid is specified by a group of three bases. These groups are called *codons*. For example, the base sequence -**CAAGTAGAC**- gets translated into the sequence of amino acids specified by the codons **CAA, GTA**, and **GAC**. The amino acids, in turn, are lined up in sequences of one hundred or more to make up different proteins. The codons cannot overlap, so -**CAAGTAGAC**- could yield the amino acids specified by codons **AAG** and **TAG**, but could not simultaneously yield those specified by **CAA, AGT** and **GTA**.

Since DNA contains four different bases, sequences of three bases could specify 4^3, or sixty-four possible amino acids. In fact, only twenty amino acids, arranged in different combinations, form all the natural proteins. Hence, many of the codons are synonymous and specify the same amino acid.

The message encoded in DNA is translated into the corresponding proteins in several steps. First the base sequence of DNA is transcribed into the base sequence of a similar kind of linear molecule, called *RNA* (*ribonucleic acid*). This process resembles the way in which the DNA strands are copied during cell division, except that only one of the two complementary strands of DNA becomes a template for the synthesis of the corresponding RNA molecule. This type of RNA is called *messenger-RNA* (*m-RNA*) because it transports the message encoded in DNA to other parts of the cell, specifically to small particles within the cell's cytoplasm called *ribosomes*, where the translation into proteins takes place.

When molecular biologists first learned how the "message" in DNA is transcribed into m-RNA, and then translated into the amino acid sequence of proteins, they were studying the metabolism of bacteria. In bacteria, this transcription happens in a regular, linear fashion from one

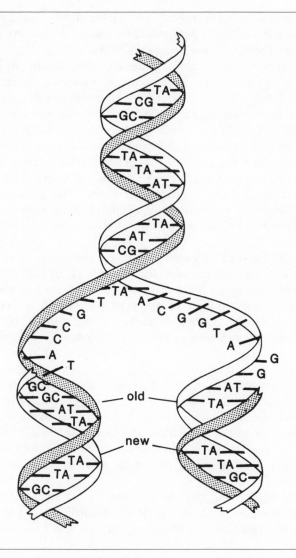

Figure 1. *Diagram of a segment of DNA.* The ribbons that form the two strands of the double helix consist of a regular, invariant sequence of phosphate and sugar molecules. The rungs connecting the ribbons are formed by the bases, which project from the ribbons toward the axis of the helix. In order for the two strands to fit together, wherever an **A** projects from one, it must meet a **T** on the other, and wherever a **C** projects from one, it must meet a **G** on the other. When DNA is copied, the double helix begins to unwind and the base sequence on each strand serves as a template for the synthesis of a new, complementary partner. (Illustration by Marie Youk-See.)

nucleotide base to the next. Molecular biologists therefore expected that, by looking at m-RNA, they would be able to deduce the sequence of bases in the DNA that was being transcribed.

As it turns out, they can make these deductions only in bacteria and viruses. In organisms whose cells contain a nucleus, the entire base sequence of DNA is first transcribed into m-RNA, but then this RNA gets modified, by eliminating parts of it and splicing the rest together into the final "mature" m-RNA, which gets translated into protein. The portions of the original DNA sequence that get edited out in this process are called *introns* and the portions that get translated into the final protein are called *exons*. In figure 2, we see schematic diagrams of three genes and, for two of them, the RNA "message" that gets translated into the corresponding proteins. Here again, we must remember that, because the figure focuses on DNA and RNA, the bulk of the story is left out.

Not only are the DNA sequences that make up one gene chopped up this way into introns and exons, but portions of one coding sequence, or exon, may function in more than one gene. Also, an intron that interrupts the coding sequence of one gene occasionally contains parts of other, unrelated genes, or even another entire gene. This is why I say that scientists create a certain amount of confusion by continuing to hang on to the concept of the gene instead of just describing how DNA functions.

Molecular biologists think that less than 10 percent of the DNA in the human genome (the totality of human genes) is coding-DNA. An even smaller fraction is thought to regulate the ways genes are drawn into the metabolic activities of cells. Molecular biologists do not know whether the bulk of the DNA has other functions, much less what those functions might be. Therefore, they refer to it as "junk DNA" or just plain "junk."

It may be that the genome represents a historical accumulation of DNA, with segments that get transcribed into RNA and translated into proteins intermingled with stretches that once had a function, but don't any more.

The process by which the message in DNA is transferred to RNA and then translated into the amino acid sequence of a protein was initially thought to operate only in that direction. It has since become clear that there are enzymes (called *reverse transcriptases*) that can transcribe RNA into DNA. Also, certain proteins can inhibit or enhance the synthesis of DNA or RNA, or affect the process in other ways. We need to think of DNA, RNA, and proteins as all acting upon one another, rather than assuming a neat line from DNA to protein.

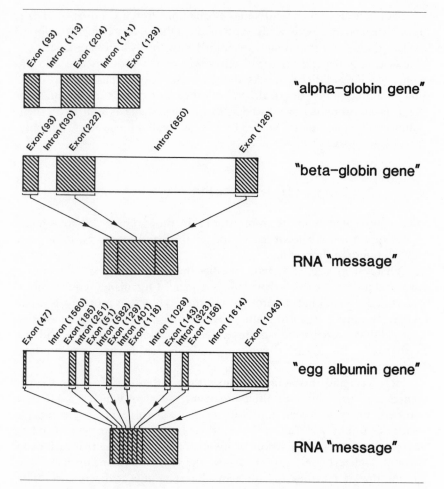

FIGURE 2. *Schematic diagrams of the genes involved in the synthesis of the alpha- and beta-globin chains of human hemoglobin, and of egg albumin.* The DNA that specifies the amino acid sequence of the alpha-chain of human globin consists of 426 bases, grouped into three exons, separated by two introns that contain 113 and 141 bases. The DNA that specifies the beta-chain is similar, except that its three exons, containing a total of 441 bases, are separated by one intron of 130 bases and a much longer one containing 850 bases. (The sickle-cell allele is a mutated form of this gene.)

The DNA that specifies the amino acid sequence of egg albumin is larger than the DNA that specifies the globins. It consists of eight exons and seven introns, which differ considerably in length. In each case, the introns are eliminated during the formation of messenger-RNA, so that the RNA "message" contains only the sequence of bases that specifies the amino acid sequence in the corresponding protein. (Illustration by Marie Youk-See.)

There are also other factors in the equation. Proteins consist of one or more strings of amino acids, but unlike DNA, which forms a long, helical thread, most proteins are folded into three-dimensional shapes. These shapes depend on the composition of the proteins and also on concentrations of different salts and the presence of metals (such as iron, magnesium, copper, or cobalt) and of other molecules (such as sugars or fats). How proteins function depends on their shape as well as on their amino acid sequence. So, while DNA is essential to the process, it is only part of the story.

HOW GENES FUNCTION

Genes differ widely in the amount of DNA they contain. Whereas most genes appear to have between five and ten thousand base pairs, some have several hundred thousand.

Whatever its size, each gene specifies the amino acid sequence of one protein. Indeed, that is what defines a gene. Organisms require thousands of different kinds of proteins in order to function properly, so even relatively simple organisms must have many genes. A change in a functional DNA sequence constitutes a gene mutation and usually leads to a change in the linear sequence of amino acids in the corresponding protein.

However, and this is crucial, since the synthesis of each protein requires the participation of different enzymes, each of which is a different protein, many different kinds of genes are involved in the complete synthesis of any particular protein. The one-to-one correspondence between genes and proteins, commonly expressed by saying that each gene "codes," "determines," or "mediates" the synthesis of one protein, only means that it specifies that protein's linear amino acid sequence. The whole process by which that protein is synthesized will only occur if the cell's entire metabolic apparatus functions properly. This always requires many different proteins and therefore many different genes.

Proteins and genes have a sort of chicken and egg relationship. Many genes are implicated in the synthesis of any given protein, and many proteins are involved in the synthesis and functioning of any given gene. Let us look at the process more closely, taking the synthesis of hemoglobin as an example. Hemoglobin is the major constituent of the red blood cells in our bloodstream, and transports oxygen from our lungs to our tissues. Hemoglobin is composed of a colorless protein called *globin*, combined with *heme*, the pigment that makes blood red. Globin contains two kinds of subunits, alpha- and beta-globin, each specified by a different gene (see figure 2). When scientists speak of globin genes, they are talking about the sequence of bases in DNA that specifies the linear sequence of the amino acids in alpha- and beta-globin.

Yet the entire blood-forming system must function properly for hemoglobin to be made. This requires not only an array of genes and the enzymes (proteins) whose amino acid sequences those genes specify but many other kinds of molecules, as well as a constellation of other requisite circumstances. The globin genes do not "cause" or "determine" the synthesis of hemoglobin. They only specify that, if the organism is functioning properly, the globins it synthesizes will have the proper amino acid sequence.

I am belaboring this point because it is usually ignored. When scientists talk about genes "for" this or that molecule, trait, or disease they are being fanciful. They attribute excessive control and power to genes and DNA, rather than seeing them as part of the overall functioning of cells and organisms.

Recent experiments that explore the way genes function have brought a number of surprises. For example, scientists now believe that the same allele of a gene can exert different effects depending on whether it has been contributed by the mother or the father. This may be the reason why some people develop Huntington disease early in life, whereas others do not develop it until their middle years.

The size of an allele may also make a difference. According to a recent study, people with myotonic dystrophy, the most common form of muscular dystrophy, seem to have a large number of repeats of the base sequence **CTG** in the relevant allele. Apparently, the number of repeats increases from generation to generation, and the symptoms of the disease become more severe the more often the sequence is repeated.[7]

At present, all that molecular biologists can do is try to correlate changes in specific genes with differences in traits. There is little by way of theory by which they could predict how, or whether, a certain mutation in a gene will affect a cell or organism. Even something as basic as the number of chromosomes or the amount of DNA an organism has in the nucleus of each cell does not tell us much about that organism's complexity or about relationships between different kinds of organisms. For example, whereas humans have forty-six chromosomes (two sets of twenty-three) in their cell nuclei, cattle have sixty, dogs and chickens both have seventy-eight and carp have one hundred and four. Nor does the number of chromosomes an organism has correspond to the amount of DNA in its nuclei. Frogs, which have only twenty-six chromosomes, have a lot more DNA than humans have.

Such numbers do not explain anything. They merely show that these characteristics do not correspond in any straightforward manner to the way organisms look or function. Scientists and physicians need to use extreme caution when it comes to making genetic predictions. DNA and its functional units play a crucial role, but a limited one. Many things that have nothing to do with genes affect the ways we develop and function day by day.

SEQUENCING THE HUMAN GENOME

As we saw earlier, the belief that genes cause traits in straightforward, predictable ways has encouraged molecular biologists to undertake the gigantic project of determining a base sequence for the DNA in all twenty-three human chromosomes. The Human Genome Project is intended to produce first a map of DNA "markers" associated with specific traits and eventually a complete sequence of nucleotide bases for a "human prototype," which will be a composite of chromosomal regions obtained from the cells and tissues of different people.

One might ask why anyone would want to undertake the herculean task of identifying the fifty to a hundred thousand genes estimated to make up the human genome and sequencing the approximately three billion nucleotide bases of which they are composed. The simplest answer is "because it is there," and because molecular biologists will no doubt learn some interesting things while doing this science. But those reasons would not elicit the kind of funding needed for this massive project. Therefore, scientists promise that having a complete DNA sequence will enable them to diagnose, and eventually cure, large numbers of gene-linked diseases. An even more grandiose reply is offered by James Watson and a number of other molecular biologists. They say that this will at last tell us what it means to be human.[8] Both of these claims are firmly grounded in the reductionist assumptions that genes cause traits and that the more we learn about their composition, the more we will know about how organisms function.

Neither of these assumptions is justified. The relationship between genes and traits is more complicated than that. This is why it does not make sense to develop the complete base sequence of a "prototype" of the human genome. Such a sequence would offer little information about the relationships between anatomical or physiological characteristics and specific genes. To obtain that kind of information, it would be far better to compare the composition of the DNA sequence of one or more specific genes in many different individuals. This would make it possible to figure out which correlations between base sequences and traits are significant and which ones are purely coincidental.

The only way to do this sort of analytical work with human DNA is to look at differences in specific base sequences among different people and try to find out which are consistently linked with specific traits. The few studies in which scientists have begun to do this show that, within the same gene, base sequences can vary a great deal without any change being apparent in the corresponding trait.

Take just one example: An international team of scientists recently reviewed all the published data collected in Europe, North America, and Japan from people diagnosed with *hemophilia B*. In people who have this condition, one of the proteins required for their blood to clot properly does not function as it should. The composition of the clotting factor and of the gene involved in its synthesis have been established, and the gene is known to contain 33,000 bases, grouped within eight exons.

The scientists found that among 216 people studied, all of whom had hemophilia B, the mutations in the relevant gene occurred at 115 different positions along the DNA.[9] This means that at least 115 changes in the base sequence can give indistinguishable results. With such a range of variability, it will be impossible to know what significance should be attached to any specific base sequence until we have looked at it in a lot of different people. This makes a mockery of the notion that one can construct one meaningful prototype for the human genome.

The scientific significance of sequencing the human genome is as questionable as the scientific significance of putting a man on the moon, but it has a similar, heroic appeal. The problem is that, quite aside from the waste of money and scientific personnel, the human genome project will have unfortunate practical and ideological consequences. Though it may not explain what genes "do," it will magnify the mythic importance our culture assigns to genes and heredity.

RFLPs: Linking DNA Patterns with Traits

Of course, not all molecular biologists are working on mapping and sequencing the human genome. Many are concentrating on specific genes or groups of them, and trying to find out how they are linked to a particular trait or health condition. In the past, geneticists have been able to put together *linkage maps* that identify the relative positions of genes on the chromosomes by virtue of the fact that certain traits seem to travel together. That is, if two traits are almost always inherited together, it is assumed that the corresponding genes lie near each other on the same chromosome, and the more consistently the traits appear together, the closer the genes are thought to lie. This has provided a rough genetic map, which molecular biologists are now trying to refine.

To identify an allele associated with a trait (say, Huntington disease), geneticists look at the DNA of people who have the condition and try to find out how its base sequence differs from the base sequence of that gene in people who do not have the condition. Given that the entire human genome contains about three billion base pairs, and that the functional base sequences represent an estimated one hundred thousand genes, that is not easy.

To find such genetic needles in the genomic haystack, molecular biologists begin by unraveling the double-helical chromosomes into single strands of DNA. They then take advantage of the existence of a family of bacterial enzymes, called *restriction enzymes*. There are many different kinds of restriction enzymes, each of which cuts a DNA strand wherever it encounters a specific sequence of about four to six bases.

In this way the restriction enzymes can be used to reduce the chromosomes to smaller pieces of DNA, called *restriction fragments*. The fragments are placed in a gel and subjected to an electric field, which makes them line up by size. Using this process on the DNA of any one person, scientists can establish a pattern of restriction fragments that is typical of that person. Because the breaks happen irrespective of whether a stretch of DNA has functional significance, the pattern of fragments may not tell us anything about that person's genes, but it will be a unique characteristic, like a fingerprint. (We will have more to say about this in chapter 11, when we look at DNA-based identification systems.)

Since mutations result in changes in the base sequence, they are likely to change the way in which the restriction enzymes cut the DNA. Therefore, a mutation is likely to result in a change in the number and size of restriction fragments. These variations in the number and length of restriction fragments are called *restriction fragment length polymorphisms*, or *RFLPs* (a "polymorph" is something that can have a variety of shapes or colors). Since different mutations change the base sequence in different ways, they give rise to different patterns of RFLPs.

The convenient thing about this technique is that one need know nothing about the genes that are actually involved with a particular trait. All one has to do is to see whether one can find a difference between the RFLP patterns of people who manifest the trait and those who do not. But there is a problem, and that is that there are a lot of variations in the base sequences of the DNA of different people. So, if someone looks at your and my patterns of RFLPs, they will find lots of differences between us. If I have a particular trait that is thought to be inherited and you don't, there will be no way to know which, if any, of the differences between our RFLPs have something to do with that trait.

The only way RFLPs can be used to establish meaningful correlations is to work within a group of individuals who are genetically related to one another—a large family or a highly inbred population—some of whom manifest the condition or trait in question and some of whom do not. Then one can try to see whether there are statistically significant correlations between a particular RFLP pattern and the trait. The question becomes: can a particular restriction fragment or group of fragments be obtained from all the people who have the trait, but not from anyone who does not?

This test turns out to be very sensitive to small changes. For example, you may have found a correlation that seems to be statistically significant. Everyone you have looked at who exibits the trait you are studying has the same RFLP pattern. Suddenly, uncle Joe shows up. He does not have the trait, but does have the pattern you thought went with it. With that, your correlation goes out the window and you need to start over again.

A few years ago, this fate befell a gene thought to be correlated with a type of manic-depressive disorder that occurs among members of a large extended family of Old Order Amish. This "manic depression gene" got a lot of publicity for a while, then one day it was gone. Two people whose RFLPs had been included among those of "healthy" family members developed the condition. In addition, the analysis was extended to several more family members who failed to fit the pattern. With that, the ten thousand-to-one probability that this disorder was related to a specific region on chromosome 11 vanished and, in fact, turned into a thousand-to-one probability against this association.[10]

To have one's conclusions contradicted by new data is a normal part of the scientific process. After all, one cannot guarantee that all of one's results will stand the test of future research. However, with the current level of interest in genes and their relationships to human diseases, each new "discovery" gets a great deal of publicity. With complex traits, such as those I will discuss in the next few chapters, all of the research is still in its early stages, so the publicity is likely to be misleading.

. . . .

FIVE

GENES IN CONTEXT

DEFINITIONS OF HEALTH AND DISEASE

Before we consider the relationships between genes and inherited conditions in more detail, we need to take a closer look at what we mean by health and illness. Our culture tends to regard these as biological phenomena, but our health is not simply a matter of biology. Social and economic circumstances affect our body states and also shape the ways we perceive and categorize them. Biology cannot be separated from social and economic realities, because they are intertwined in complex ways and build upon each other. We cannot isolate the biological factors, and when we try we oversimplify and distort reality.

The categories of health and illness describe a continuum of states. At one end we feel on top of things, at the other we feel terrible, and there are many gradations in between. Where along that line we decide we are sick, at what point we decide we need advice from others, and who those others are depend on cultural practices and the costs of medical and other assistance.

If we go to an expert in "bad-feeling," a physician or nurse practitioner, we run into a phenomenon that is typical of our medical culture: there is often a difference between the way we feel and our medical "diagnosis." Medicine generalizes individual people's feelings and fits them into ready-made diagnostic categories, as though diseases were discrete entities with an existence of their own. It encourages scientists to search for definable causes and cures without raising questions about how diseases fit into the life circumstances of specific individuals.

The notion that diseases are not attributes of specific individuals, but can be named and characterized as though they were independent of our lives and feelings, evoked vigorous disagreements and debates well into the nineteenth century. Turn-of-the-century writers like Tolstoy and Rilke complained that the new, scientific medicine was severing our existential relationships to our diseases and therefore to ourselves. Tolstoy's Ivan Ilyich complained that his physician was ignoring him and was only interested in his kidneys.[1] Rilke's Malte Laurids Brigge complained that hospitals make it impossible for people to die their own personal, albeit agonizing, deaths.[2]

The idea of disease as a diagnosable essence did not enter our culture without a fight, but enter it did. Now, though we complain when we take our feelings of malaise to the doctor and he or she tells us that "nothing is wrong," we accept the verdict. Similarly, we have learned to consult medical practitioners for so-called checkups, even when we feel fine. At such times we are basically asking them to tell us whether it is legitimate for us to feel all right. Whether we are diagnosed as healthy or ill, we tend to assign more value to this judgment than to our own feelings. In fact, sometimes a diagnosis of "nothing wrong" makes us feel better than we did before and a diagnosis that something is wrong can make us feel bad even though nothing else has changed.

In the case of infectious diseases, the search for unique causes has helped scientists to identify bacteria and viruses and to study their transmission, so that public health or medical measures can be taken to minimize or prevent their occurrence and to treat them with antibacterial or antiviral drugs or vaccines. However, as the political scientist Sylvia Tesh has shown, even here the picture is far from simple and requires a good many political assumptions.[3]

Contrary to what we are usually taught in school, mortality rates from almost all known infectious diseases were already decreasing in the industrialized world many decades before the offending bacterial or viral agents were identified. Deaths from such serious scourges as tuberculosis, scarlet fever, measles, and whooping cough were on the decline long before the vaccines or drugs that are effective against these diseases were developed.[4] Thomas McKeown, a British population scientist, attributes this decline to innovations in agriculture and transportation that increased the availability of different foods and so improved nutrition, and to sanitary measures that provided more healthful water and better sewage disposal and housing.

The other side of the coin is that access to health care is not enough to make people healthy. Data collected in Great Britain in the 1970s (even before the Conservatives began to dismantle the National Health Service) show that, while twenty-five years of universal access to free health care did improve people's health, mortality rates were still inversely correlated with social class. The lower a family's social class, the higher the mortality rate of both women and men of that class. This was equally true for treatable infectious diseases, for chronic conditions such as cancer or diseases of the circulatory or digestive system, and for causes of death that clearly are socially mediated, such as accidents, poisoning, and violence.[5]

In the United States, we are now seeing epidemics of several major infectious diseases that we had thought long gone, or at least under control, among them tuberculosis, measles, syphilis, and gonorrhea.[6] Many of these epidemics can be traced to specific social factors, such as overcrowding and lack of funds for vaccinations and other public health measures. However, biomedical researchers tend to concentrate on problem solving, rather than on ways to make existing solutions accessible. By focusing our attention on microorganisms or genes, scientists succeed in drawing attention away from societal influences. They also ensure their own monopoly, by keeping disease prevention in scientific institutes and laboratories.

Health and disease thus get defined as scientific problems for which we must seek scientific answers. As Tesh points out, medical scientists create the illusion that health is a technical problem, rather than a social problem that requires social remedies at least as much as medical ones. They promise to overcome the mysteries involved in chronic conditions, just as they have mastered infectious diseases. They dismiss hypotheses about the social origins of disease as unscientific and "political."[7]

INDIVIDUALIZATION OF HEALTH AND ILLNESS

While medical scientists often reject social explanations of biological states, they can be more accepting of biological studies that purport to explain social conditions. For instance, Daniel Koshland, a molecular biologist and the editor of *Science* magazine, prophesies that the Human Genome Project will "aid the poor, the infirm, and the underprivileged," because it will improve physicians' abilities to diagnose, and presumably cure, "mental illness." In a *Science* editorial, Koshland states, as though it were an uncontested fact, that mental illness is "at the root of many current social problems" and that understanding the human genome will enable us to move beyond "the current warehousing or neglect of these people."[8]

In this analysis, Koshland not only is making false promises but is actively drawing attention away from the economic and political realities that victimize people. He is diagnosing the poor and underprivileged as sick, and touting the better understanding of genes as a cure for what are clearly economic, political, and social ills.

Koshland is giving a new twist to a very old idea: that people are poor, or rich, because of something inside them rather than because of social inequities. In the nineteenth century, when the science of genetics was born, it was commonly said that "blood will tell." Fictional heroes like Oliver Twist exhibited the virtues and honesty of their middle-class parents, despite their cruel, working-class upbringing. The merchant class attributed its success to a natural superiority over those who failed to rise from poverty. Now, geneticists have translated such perceptions into scientific terms.

In our day, it is no longer acceptable to say that the poor are genetically inferior, but Koshland seems comfortable implying that people are poor because they are mentally ill. Like the genetic explanation, this explanation medicalizes and individualizes the problems that underlie the current situation of poor and homeless people. This is a process that is familiar to all of us. When physicians and policymakers treat smoking, alcoholism, cancer, or heart disease as individual health problems, they ignore the societal and environmental factors that contribute to these conditions.

Of course, to some extent it makes sense to think about one's health in personal, hence individual, terms, since each of us is most concerned about our own health and the health of those who are dear to us. However, our state of health depends not only on what goes on inside our bodies but also on the conditions under which we live and work. Individual susceptibilities may play a part, but many of the preconditions for our health are beyond the control of any but a privileged few.

This dialectic is at the heart of the old debate between physicians who place primary emphasis on curative medicine and those who emphasize public health measures. We need both, but an excessive preoccupation with individual concerns and responsibilities is detrimental to health when it encourages us, as a society, to neglect the systemic conditions that affect all of us.

There must be a balance between public health measures and individual health care, and between the resources that need to be allotted to each. In an urban slum in Guatemala, where the dirt roads are bordered by open sewers and the only nominally clean water is dispensed by one spigot per thousand inhabitants, it makes little sense to test the susceptibilities of individual children to infectious organisms. Only when the obvious sources of infection that affect everyone have been greatly reduced does it make sense to single out the people in need of special care.

In the same way, it makes no sense to identify the susceptibilities of individual factory workers to industrial dusts or other toxic substances without reducing the hazards under which they work. Yet the drumroll

Cotton-bale opening room, Watershoals, South Carolina. The flimsy mask does not protect this worker or anyone who comes in contact with his clothing against inhaling cotton dust and fibers. (Photograph © Earl Dotter.)

of publicity that touts genes as "causes" of an ever-expanding range of traits, diseases, and disabilities draws our attention to the affected individuals and away from the conditions that provoke their problems.

These conditions, and their effects, should be clear to anyone willing to look at them. For example, in 1988 (the latest date for which this information has been compiled), the infant mortality rate for African-American babies in Boston was three times that for European-American babies. This medical Mecca, which boasts sixteen teaching hospitals, had the third highest black infant mortality rate in the United States (24.4 deaths per thousand births).[9]

Infant mortality is not only an urban problem, nor is it limited to people of color. Since 1950, when the United States had one of the lowest infant mortality rates in the world, it has fallen to around twenty-fifth place. The present U.S. infant mortality rate is higher than that of any other industrialized nation, except South Africa and Israel, and is comparable to rates in several of the poorer nations of the so-called Third World, such as Cuba and Barbados. Meanwhile, with all the medical and technological advances of the twentieth century, the United Nations

Children's Fund reports that, in the world as a whole, "more than a quarter of a million small children are dying *every week* of easily preventable diseases and malnutrition."[10]

In a somewhat different vein, a recent article in the *American Journal of Public Health* documents the existence of massive, unreported workplace exposure to lead in a variety of industries in which the levels of environmental contamination are not even routinely monitored. The affected workers are not aware that they have been exposed, and without monitoring there is no way for them to find out, despite the fact that the severe neurological damage produced by lead poisoning can be reversed if it is caught in the early stages. Not only are the contaminated workers themselves at risk but they are liable to carry the contaminants home to their families on their hair, skin, and clothing.[11]

Let us look at some other effects of this one environmental poison. A recent panel assembled by the Centers for Disease Control in Atlanta estimates that more than six million U.S. children are at risk of lead poisoning from lead paint, gasoline, and other sources, at levels likely to stunt their growth and limit their abilities to succeed in school.[12] And lead is only one among numerous environmental contaminants known to be detrimental to health. Yet in September 1990, when the U.S. secretary of health and human services, Dr. Louis W. Sullivan, enumerated the preventive measures needed to improve the health of Americans, what tops his list is the need for greater participation in exercise, physical fitness, and weight loss programs—all individual solutions most readily available to people with money and leisure. The need to reduce infant mortality is way down on his list and he does not even mention the need to monitor and decrease exposures to environmental hazards in the workplace and elsewhere.

Ironically, a newspaper story reporting Dr. Sullivan's program ran under the headline, "Health goals are set for the Year 2000; *Prevention Stressed*" (italics mine).[13] The writer of this story and, apparently, Dr. Sullivan himself seem to think that "prevention" involves mainly the behavioral choices that relatively affluent individuals are in a position to make, rather than the economic, public health, and medical measures that would reduce unnecessary risks and improve everyone's chance to be healthy.

GENES AS BLUEPRINTS

Inherited factors can have an impact on our health, but their effects are embedded in a network of biological and ecological relationships. Genes are part of the metabolic apparatus of organisms that have multiple, mutual relationships with their environments. We breathe our "environment," eat it, sweat and excrete into it, move through it and with it.

This is one reason that even "simple" Mendelian conditions exhibit varying degrees of severity. Concepts like "the organism," "the gene," or "the environment" are useful as ways to organize our understanding of the world, but we must keep in mind that they do not describe the world as it is. They merely serve to separate out the specific aspects on which we want to focus our attention.

As we have seen, genes affect our development because they specify the composition of proteins, but it is more realistic to think of genes as participating in various reactions than as controlling them. Because of their complexity and their ability to adapt to change, organisms can sometimes develop ways to compensate for the failure of specific reactions to take place, or for reactions that occur too rapidly or slowly. So, when molecular biologists speak of genes as "control centers" or "blueprints," this is testimony to the hierarchical models they use rather than a description of the ways in which organisms function.

Each protein, and therefore each gene, can affect many of an organism's traits. Conversely, each trait receives contributions from many proteins, hence from many genes. For example, when the gene that specifies the structure of human growth hormone (a protein) was transferred into a mouse embryo, the mouse grew to twice its normal size. When the same gene was inserted into the embryo of a hog, that animal's size did not change, but it became leaner than normal.

The ways this gene functioned depended on other things going on in the organism. To say that the gene "caused" the effects dodges the question of why these effects were different. Obviously, the gene played a part in both cases. Equally obviously, it was not the only factor. Molecular biologists emphasize the role of genes in this situation because they are more interested in genes than in the development of mice or hogs.

Within a single species, as well, the same gene can contribute to different effects in different individuals. In very few cases can a gene legitimately be said to be "for" any one thing. Scientists now know the precise molecular structure of the allele associated with sickle-cell anemia, and for several decades they have known the specific molecular change in sickle-cell hemoglobin that is responsible for this condition. Yet, this knowledge has not enabled them to understand why some people who have sickle-cell anemia are seriously ill from earliest childhood, while others show only mild symptoms later in life, nor has it helped produce cures or even effective treatments. The best medical therapies for people with sickle-cell anemia still rely on antibiotics that control the frequent infections that accompany the condition.

It is misleading for proponents of the genome project to promise that knowing the sequence and composition of all the genes on the human chromosomes will lead to cures for a wide range of diseases. It is all too easy to find proteins associated with specific health conditions, and with present techniques it has become possible to identify genes that specify the composition of these proteins. Such discoveries can be useful, in that they may make it possible to produce large quantities of the proteins, which will make it easier to do research on these conditions. However, this will not necessarily identify their "causes" or cure them.

Only rarely can information at the level of DNA sequences be readily translated into useful information at the level of cells, tissues, or whole organisms. In the past, scientists have deduced the presence of genes, as well as their functions, by looking at the ways organisms differ from one another. There is no reason to take it for granted that this scenario can usefully be played backwards and that now scientists will be able to identify a gene's critical function, or functions, when they have located, isolated, and sequenced that gene. "Predictive genetics" may work in a few special situations in which a particular DNA sequence points to specific and special characteristics that occur in only a few proteins, but most DNA sequences will not be that informative.

GENETICIZATION

It has become fashionable to look for genetic explanations for health and illness. The argument runs like this: Environmental factors influence many aspects of our health, but despite the fact that people who smoke are at a greater risk of getting lung cancer than those who do not, not every one who smokes gets lung cancer. Conversely, not every one who gets lung cancer smokes. So, something other than smoking distinguishes people who get lung cancer from those who do not. To scientists who consider genes to be the basis of our entire biology, genes are the likeliest culprits.

As the human genome is analyzed at a new level of detail, correlations inevitably will turn up between certain DNA sequences and particular diseases or other traits. But, until the DNA sequences of large numbers of people have been looked at, it will be impossible to distinguish significant correlations from accidental ones. Unfortunately, at this point each correlation that results in a scientific paper tends to give rise to a news headline. When later scientific papers show the correlation to be false, that sometimes rates another headline, but often it does not.[14]

Already the confusion is enormous. Within the last few years, genes have been announced "for" manic-depression, schizophrenia, alcoholism, and smoking-related lung cancer. The claims about manic-depression and schizophrenia genes were withdrawn soon after their announcement and the gene for alcoholism met the same fate later, although another one has since crept into the news. These supposed identifications are invariably obtained with small numbers of people, and much publicity accompanies every such "discovery." Although, like mirages, many of these genes disappear when one tries to look at them closely, a confusion of claims and counterclaims is inevitable, and there are so many stories that people are left with the impression that our genes control everything.

In later chapters we will look in more detail at evidence about the extent to which genes contribute to various sorts of health conditions. Right now I want to consider the way that the picture of genes as control centers has expanded the category of "genetic disease."

In addition to the relatively few and rare conditions whose patterns of inheritance can be described by Mendel's laws, scientists and physicians increasingly speak of inherited "tendencies" or "predispositions" to develop more complex and prevalent conditions. In most of these cases, they are just using the word "gene" as shorthand for their belief that the condition is inherited biologically, even though they cannot be sure that it is and cannot predict who will inherit it. Complex conditions are variable and unpredictable, and involve a wide range of biological and environmental factors. It is not clear that identifying genes will give us a better picture of what is going on.

Considering the variety of social and economic risks all of us face, it seems a distraction from our obvious, daily problems to focus on the risks we may harbor in our genes. Worse yet is the implication that it would be irresponsible to go on living without this knowledge, even though there is little we can do once we have it. Yes, we can eat more healthful foods, but only if we can afford to buy them. And yes, we can decide to stop smoking and drinking, but only if the circumstances of our lives make such changes possible. Anyway, these changes would be good for all of us, irrespective of our genetic "predispositions." The unwarranted individualization of responsibility for our own health and that of our children and the fatalism genetic tests can engender may, in fact, prevent some of us from doing things we might otherwise do to stay healthy.

In a recent article, published in the scientific journal *Genome*, the medical geneticist Arno Motulsky promises that "definite prediction of somatic [that is, physical] and some psychiatric disease will be increasingly possible in the future." Yet, in the very next sentence he points out that "in many conditions predictions will not be 100 percent accurate," but will only mean "that a given disease will occur with a greater statistical likelihood than expected in the general population."[15] This is an odd kind of "definite prediction." It does precisely what Sylvia Tesh suggests scientists do when they try to appropriate our health: It promises major benefits that in reality add up to little.

Despite the scientific problems that surround the identification of genes "for" specific conditions, and the social and personal problems such predictions can entail, our current infatuation with genetics pushes genes into the foreground. Both the scientific pronouncements and the ways they are reported in the press often imply that, with a snap of their fingers, scientists will progress from the point at which they have identified a gene they suspect may be associated with some devastating condition such as cancer, to predicting whether an individual will develop the condition and, better yet, to curing or preventing it.

Let us look at an example: In September 1990 an Associated Press story announced that "Scientists . . . have cloned a gene that helps brain cells communicate, a step that may lead to improved drugs for schizophrenia and . . . may someday help doctors diagnose schizophrenia and Parkinson's disease before symptoms appear."[16] This is the sort of thing scientists and science writers say to stimulate interest. We have all heard of schizophrenia and Parkinson's disease, and cloning "a gene that helps brain cells communicate" sounds impressive. The claim may be true, but the reality is much more complicated than they suggest. Scientists have identified a DNA sequence, implicated in the synthesis of a protein called the *dopamine receptor*, which occurs in brain cells. Dopamine is one of several small molecules that are released by some nerve cells in the brain and taken up by others. This is what is meant by the word "communicate." Scientists do not know what these brain cells say to each other, nor exactly how they say whatever it is that they do communicate. They just know that dopamine and other neurotransmitters are involved.

Dopamine is also thought to be involved in some way in Parkinson's disease, since the tremors and other symptoms of some, though not all, people who have this condition are reduced when they take dopamine or compounds chemically related to it. Again, why it works is not understood, though the assumption is that, by binding and releasing dopamine, dopamine receptors may modulate the concentration and activity of this chemical in the brain. If that is so, the gene that specifies the receptor's structure may affect dopamine activity. The Associated Press story quotes scientists who suggest that once this gene has been isolated, they may be able to study the dopamine receptor in greater detail than was possible before, and that they may then be able to develop drugs that modify the receptor's interactions with dopamine.

In other words, scientists intend to use the gene as a biochemical tool to study the metabolism of dopamine and try to develop drugs that mimic or counteract its action. This is a reasonable plan of biochemical experimentation, but what makes it interesting to the public is its association with familiar diseases and stories about how the brain works. Molecular biologists are also attracted by these associations. They like to feel that they are getting at the root of human thought and action. We are back to reductionism: Brain function gets explained in terms of the activity of molecules in the brain, hence of the genes that participate in the synthesis of these molecules. Then, how people act gets explained in terms of brain function, hence of such molecules and genes.

We need to understand the patterns that underlie the grandiose scientific announcements and the ways they are reported in the press, because more and more of them are being made as it becomes technically easier for scientists to isolate genes and produce them in quantity. A host of therapeutic claims has been spawned by the idea that identifying a DNA sequence and the protein whose composition it specifies will lead to cures for a condition associated with that protein. Whenever a DNA sequence is isolated that specifies the composition of a protein involved with the ability of cells to multiply or stick together, scientists say they are on the road to curing cancer. Locating a gene that specifies a protein involved with cholesterol metabolism puts them on the road to conquering high blood pressure, strokes, and heart disease. And so on. But metabolic relationships and their derangements are too complex to permit such simplistic solutions.

Diagnostic Labeling

There is an even more pressing argument against research that seeks to identify genes "for" this or that condition. The development of tests to detect genes, or substances whose metabolism they affect, opens the door for the invention of an unlimited number of new disabilities and diseases. For any metabolite or other trait that has a normal distribution in the population, some people can be defined as having "too much" and others "not enough." (In mathematical terms, *normal distribution* simply means that most people cluster around some average value that gradually falls off toward zero on both sides of that average or *mean*. "Normal" in the colloquial sense means whatever the society wants it to mean.) Pharmaceutical companies and physicians stand to make a good deal of money from inventing new diseases as fast as new diagnostic tools are developed that can spot or predict their occurrence.

Let us look at an example. Genentech, one of the first generation of biotechnology firms, markets a genetically engineered form of human growth hormone. This hormone previously could be obtained only in minute amounts, by isolating it from the pituitary glands of human cadavers. When the supply was limited, human growth hormone was only used to treat children with *pituitary dwarfism*, which results from the reduced secretion of this hormone by the pituitary gland. Once the hormone became available in quantity, physicians began to prescribe it to treat people who secrete normal amounts of growth hormone.

In one series of experiments, growth hormone was given to growing boys deemed "too short" for their age. A *New York Times Magazine* cover story on these experiments reports that Genentech scientists have suggested that it is proper to consider any child whose height falls within the lowest 3 percent of the population as suitable for treatment.[17] But it is in the nature of characteristics like height that, no matter what their average distribution may be, there will always be a lowest—and highest—3, or 5, or 10 percent. Physician John Lantos and his colleagues point out that "of the three million children born in the U.S. annually, 90,000 will, by definition, be below the third percentile for height." This "treatment" is not without risks. There is no telling how the health of these children will be affected by daily injections of growth hormone. However, since growth hormone treatment costs about $20,000 a year per child, if each of these children received a five-year course of treatment this would constitute a potential market of about nine billion dollars a year for Genentech.[18]

Height is not the only characteristic for which people are using growth hormone. Recently rumors have been circulating that athletes are using it to build up their muscles. Since the level of growth hormone varies from person to person, artificial supplements of it would be harder to detect than the metabolic steroids some athletes have used for this purpose. But human growth hormone, in excess, is by no means harmless. People whose pituitary gland secretes too much growth hormone often develop *acromegaly*, a condition that involves an overgrowth of the bones of the hands, feet, and face. Thus the use of this hormone to "treat" healthy people seems hardly justified.

Researchers have also suggested that administering growth hormone to old people could slow the aging process. A report of the use of synthetic human growth hormone for this purpose appeared in July 1990. The experiment, published by ten physicians in Milwaukee and Chicago, involved twenty-one men between sixty-one and eighty-one years old.[19] These men reported no symptoms and were selected as subjects merely because, on two successive measurements, their hormone levels were in the lowest third of the normal range. Twelve of them were given sufficient amounts of human growth hormone to bring their levels into the range found in "healthy young adults." The other nine served as "controls."

Since all these men were healthy to begin with, the benefits of the treatment were measured by the following "symptoms": mass of fatty tissue, which tends to increase with age; overall muscle mass, which tends to decrease; and skin thickness and bone density, both of which tend to diminish. The experiment showed that, at a cost of about $14,000 a year, these indices could be brought into a more "youthful" range. However, the author of an accompanying editorial points out that long-term administration of growth hormone can elicit diabetes, arthritis, hypertension, edema, and congestive heart failure.[20] Perhaps a more fundamental question is whether the fact that human growth hormone can now be produced in quantity justifies turning the normal process of aging into a disease.

Stories like these demonstrate that in a capitalist economy it is virtually impossible to develop products that can benefit only a few people. Once such a product becomes available, and especially if it has been expensive to produce, the producers will do what they can to expand the market for it, even when its wide use poses known dangers.

The market for artificial "improvements" will simply depend on where one decides to draw the line for what is to be labeled "abnormal." This is true for any numerical trait—height, weight, amount of body fat, metabolic rate, and so on. Now that biotechnology companies are producing growth hormone, an obvious next step is to produce an "anti-growth hormone" that promises to slow growth. Perhaps the companies could market it to parents of children, especially girls, who are predicted to be among the tallest 3 percent of the population.

If a boy is "too short" or a girl "too tall," if a woman's breasts are "too large" or "too small," if a man wishes he had been born a woman, or a woman that she were a man, they need only find a physician who can administer the right substance and their troubles will be over. Except that their troubles—and ours—will have just begun. There will always be people who would like to change their children or themselves and novel medical treatments won't cure such insecurities. As long as every deviation from the standard, prepackaged norm is considered "abnormal," physicians, geneticists, and the biotechnology companies will not run out of customers.

SIX

. . . .

INHERITED "TENDENCIES":
CHRONIC CONDITIONS

SOME UNDERLYING ASSUMPTIONS

Medical attempts to discern the roots of future diseases in healthy individuals did not begin with the introduction of genetic tests. In the late 1960s and early 1970s the Kaiser-Permanente group and some other health maintenance organizations (HMOs) tried to institute "multiphasic screening programs." As part of their medical checkups, healthy clients received batteries of medical tests, the results of which could be fed into computers in order to generate predictive "disease-profiles." The expected advantage of such testing was that physicians would be able to institute preventive measures that would keep people healthy, instead of waiting to treat them after they got sick. This was meant to benefit the HMO as much as the clients, since it was assumed that such tests would not only prevent diseases, but would be cheaper to administer than treatments.

These expectations proved wrong. Since even the best tests produce significant numbers of false results, it soon became clear that administering many different tests to tens of thousands of people inevitably generates large numbers of faulty diagnoses. As a result, even if the number of mistaken diagnoses on any one test is as low as 1 percent, many people will be treated inappropriately, which is dangerous for them and expensive for the HMO. Despite these problems, predictive diagnoses are again coming into vogue now that scientists and biotechnology firms are ready to produce predictive genetic tests.

When they are correct, predictions and early diagnoses may appear to be helpful, but unfortunately this is often an illusion. For example, when we are told that preventive screening and early diagnosis has prolonged the survival of women diagnosed to have breast cancer, we must understand that this does not necessarily mean that they live longer than they would have lived otherwise. An early diagnosis, by definition, increases the length of time a person survives after the condition has been diagnosed, whether treatment helps or not. For this reason, publicity about presymptomatic screening and early detection increasing survival times needs to be looked at carefully before we accept this as a benefit of predictive testing.

To expose large numbers of people to diagnostic tests is always risky. Individual susceptibilities vary and even relatively low-risk procedures are likely to be hazardous to some people. For example, researchers at the University of North Carolina suggest that mammograms may induce breast cancer in many more women than has been supposed. They have found that about 1 percent of Euro-Americans apparently inherit a susceptibility to radiation that puts them at risk from even the relatively low dose of x-rays delivered by mammograms.[1]

Since there is no way to identify people who have this increased susceptibility, the researchers suggest that it may be safer for women to have thorough, periodic manual examinations than to have mammograms. But what if a test were developed to identify women who carry the gene that is supposed to confer this increased susceptibility to radiation? Would physicians then begin to screen women so as to detect the ones who may be put at risk by a further screening test, to wit, a mammogram? If we are going to institute genetic screening for susceptibilities to the potentially adverse effects of screening tests intended to detect susceptibility to cancer or other conditions, where do we stop?

Some people ask, "But what if I really do have a genetic tendency to develop a condition, such as high blood pressure? Wouldn't it help me to know that?" Yes and no. For one thing, if you have this or any other "tendency," it does not mean you will develop the condition. And if you do not have that "tendency," it does not mean you won't develop it. It is best to make dietary and other adjustments that reduce the likelihood of developing high blood pressure, irrespective of whether you are told you have such a "tendency." If you cannot or do not want to make these adjustments, the discovery of an inherited "tendency" is not likely to help.

Lately, medical experts have begun to question whether it is worth trying to bring the blood pressure and cholesterol levels of healthy individuals down into the "normal" range by making them follow a strict diet, get lots of exercise, and sometimes even take blood pressure-lowering drugs. This discussion has been provoked by the publication in the *Journal of the American Medical Association* of a long-term Finnish study of 1,222 male business executives between the ages of forty and fifty-three who were physically well but had risk factors for developing heart disease.[2] Of these men, 612 were medically supervised, while the others served as controls. To their surprise, the scientists conducting the study found 2.4 times the rate of deaths from heart attacks, and 45 percent more deaths from all causes in the group that received the "preventive" treatments than among the controls.

A U.S. physician, asked to comment by the *Boston Globe*, is quoted as saying: "One disadvantage of a mass prevention strategy is that in order to save a few or help a few, you're subjecting a large number of people to drugs or other interventions."[3] That's one important point. Another, as I have said before, is that for any numerical trait—height, weight, blood pressure, cholesterol levels—what counts as "normal" is a matter of definition. Perhaps the "high" cholesterol levels and blood pressures were normal for some of the men in the Finnish study.

Even where preventive measures may be helpful, the circumstances of our lives, including our income, work, and family situation, may make it difficult or impossible to "follow doctor's orders." It does not help us to be told that we have this or that "tendency," if there is nothing we can do with this information. Medical, and especially genetic, predictions do not increase individual control. Such predictions place the source of all our health conditions in our biology and give physicians and scientists authority over them. By erasing the social context, genetic predictions and labels individualize our problems, blame the victim ("If you get sick, it's because you have bad genes"), and are authoritarian ("You should have had your genes tested and done what the doctor said!"). But predictive tests contain rather little information to live by, since the answers they offer are almost always couched in terms of probabilities and contingent on other factors.

Conditions That Run in Families

The manifestations of inherited conditions can vary considerably and unpredictably from one individual to another. The reason is that many factors, both within and outside ourselves, affect the ways we develop and function. This is true of our biological characteristics as well as of our psychological and social ones. Even conditions such as Huntington disease, cystic fibrosis, and sickle-cell anemia, which follow predictable patterns of inheritance, can exhibit a wide range of symptoms that differ in their severity among different people.

Other conditions, which show no predictable patterns of inheritance but seem to "run in families," increasingly are also being attributed to genes. Before we look at specific examples, let us remember the case of pellagra, the nutritional deficiency produced by a lack of nicotinamide, which was thought to be genetic since the lack of this vitamin did, indeed, run in poor families in the southern United States (see chapter 2). By the same reasoning, in the mid-nineteenth century, Francis Galton, the father of eugenics, attributed both the superior intellectual and professional achievements of successive generations of upper-class Englishmen and the deplorable habits of "paupers" and "criminals" to their inherited biological constitutions.[4]

Nowadays we tend to hear less about the biological transmission of criminality. (Although in 1992, researchers at the University of Maryland, with financial support from the Human Genome Project, announced a conference on "Genetic Factors in Crime." But, following widespread objections, funding was withdrawn.[5]) We do hear a good deal about the possibility that alcoholism and other addictions are inherited biologically, and these conditions are then blamed for a panoply of socially unacceptable behaviors.

We need not be surprised that molecular biologists can establish links between genes and complex health conditions or behaviors. Of course genes participate in such processes, since DNA specifies the amino acid sequence of the proteins involved in all biological functions. However, such "genetic" components will be at most contributing factors, and we should not expect the conditions with which they are associated to exhibit predictable patterns of inheritance.

All biological traits are what genetic jargon would describe as "polygenic and multifactorial"—they involve many genes and processes that take place in and outside the organism. The present interest in deciphering these processes is based on the belief that the transmission and development of any trait can be modeled by a hierarchy of "causes" presided over by the gene. Medical scientists like this model because it promises that once "the gene" has been identified, they will be able to diagnose the condition more readily and learn to understand its molecular basis and possibly cure it. Failing that, they might at least be able to predict the occurrence of a condition before its symptoms become evident. In that case, they can try to delay or prevent its manifestations.

In this chapter and the next, we shall look at examples of conditions that sometimes "run in families" and at ways scientists hope to identify DNA sequences that could serve as predictive genetic "markers" for them. This is a questionable project because human populations are quite variable. A DNA sequence that is a marker for a particular trait in one family may not be associated with that trait in another. Therefore such markers will tend to elicit needless fears or, alternatively, to offer unwarranted reassurances.

DIABETES

Diabetes is a disturbance of carbohydrate metabolism, characterized by unusually high concentrations of the sugar *glucose* in the blood. Two to 4 percent of people in the industrialized countries and about 0.1 to 1 percent in poorer countries have diabetes.[6] Medical scientists recognize two forms of diabetes, called type 1 and type 2. About 10 percent of people who have diabetes have the type 1 condition. In Scandinavia, where the incidence is highest, this is around four out of every thousand people. Type 1 diabetes usually appears during adolescence, though it can start earlier or later, and it begins quite suddenly. By contrast, type 2 diabetes tends to come on gradually and not until people have passed their middle years. Often, people who develop type 2 diabetes are considerably heavier than average for their height.

The metabolic patterns underlying the two forms of diabetes are quite different. Type 1 diabetes results from the destruction of cells in the pancreas that normally produce insulin, a hormone involved in sugar metabolism. Type 1 diabetes is thought to involve the immune system and be the result of an allergic response to toxic chemicals in the environment, a viral infection, or some other unidentified stimulus. Since people who have this form of diabetes stop being able to produce insulin, which is essential for carbohydrate metabolism, they must regularly receive appropriate amounts of this hormone.

By contrast, people with type 2 diabetes secrete normal or above-normal amounts of insulin, but their tissues develop an insensitivity to it. Therefore the insulin loses its metabolic effectiveness. Type 2 diabetes, which is by far the more common of the two forms, can often be alleviated by a diet low in carbohydrates and fats, especially when coupled with moderate levels of exercise. Indeed, a study of nearly 6,000 middle-aged men, published in the *New England Journal of Medicine*, showed that regular exercise such as jogging, bicycling, and swimming markedly reduced the incidence of type 2 diabetes.[7]

Reporting on this study, Dolores Kong writes in the *Boston Globe* that the greatest "protective benefit of exercise showed up in men who were at the greatest risk because of obesity or who had a family history of diabetes." She quotes a Dr. Evans at the U.S. Department of Agriculture's Human Nutrition Research Center at Tufts University, who says that "for many, many years, we've thought that this kind of insulin insensitivity was kind of an inevitable consequence of advancing age. [But now] it appears that it is nothing more than an accumulation of a lifetime of inactivity. That means the process is very reversible."[8] While this does not rule out the possibility that type 2 diabetes has a genetic component, other factors clearly play an important role.

Molecular biologists believe that several proteins are involved in the development of type 2 diabetes. Among them are insulin and the "insulin receptor," a protein on cell surfaces with which insulin needs to combine before it can exert its metabolic effects. At present, the hunt is on to locate, identify, and sequence the genes that specify the amino acid sequences of insulin and the insulin receptor.

Molecular biologists think that the "insulin gene" lies on chromosome 11 and the gene involved in the synthesis of the insulin receptor lies on chromosome 19, though there could be more than one receptor protein. Once enough is known about the structure and location of these two genes, scientists will be able to develop tests to detect differences in their base sequences. Such tests could then be used to predict a "predisposition" to develop type 2 diabetes in healthy people who are members of families in which the condition occurs.

However, not every difference in the base sequences of these genes will result in significant changes in the ways the proteins function, and there is no way to know in advance which ones matter. To date, scientists have identified several mutations in both the "insulin gene" and the "receptor gene," but none of them appears to be correlated with the incidence of type 2 diabetes.[9]

All of this research is being done in the hope of finding a predictive test for a "predisposition" to develop a condition that many people could avoid by changing their diets and getting regular exercise. Surely, it would be better to educate everyone about the importance of diet and exercise and to work toward providing the economic and social conditions that could enable more people to live healthily, rather than spending time and money trying to find "aberrant" alleles and to identify individuals whose genetic constitution may (but then again, may not) put them at special risk.

Meanwhile, what is happening with type 1 diabetes? The susceptibility to type 1 diabetes appears to cluster in families and in specific populations, for example among people of northern European origin. If one child in a family has type 1 diabetes, the probability of a sibling developing it is about 6 percent, or twenty times the rate for the general population. While this might seem to indicate a genetic component, it turns out that an identical twin of someone who develops type 1 diabetes has only a 36 percent probability of developing the condition.[10] This is higher than the probability for ordinary siblings, but proves that genes cannot be the sole determining factor. Anyway, since toxic environmental agents and viral infections are thought to provoke type 1 diabetes, family correlations need not point to a genetic origin. Siblings who live together are often exposed to the same environmental agents.

Nonetheless, molecular biologists are trying to develop predictive genetic tests for this condition. This time they are not looking at the "insulin gene," but at genes that participate in the synthesis of proteins active in immune reactions, tentatively localized on chromosome 6. Whatever they find, we can be sure that predictive diagnoses will be tentative at best, both because of the complexities of the immune system and because no one knows what factors trigger this particular immune response.

I am profoundly skeptical about this line of research. It is not useful or, indeed, possible to translate the complex relationships among the various factors that participate in metabolism, growth, and development into hierarchies of causes, with genes at the top. However, what concerns me is not just that such projects of diagnosis and therapy may fail. Failures are an integral part of research, and are often educational. The problem is that, despite the practical failures, the projects will succeed in bolstering the genetic ideology. Especially when linked to a condition such as diabetes, whose incidence probably could be reduced by education and public health measures, such beliefs support the neglect endemic in our contemporary politics and health care system.

HIGH BLOOD PRESSURE, HEART DISEASE, AND STROKES

Large numbers of people in the industrialized parts of the world suffer from conditions linked to the build-up of fatty substances inside the walls of blood vessels or on the valves that guide the flow of blood within the heart. The collective medical term for this set of conditions is *atherosclerosis*. (*Ather* refers to the fatty deposits and *sclerosis* to the decreased flexibility of the walls of the blood vessels or of the heart valves.) The excessive rigidity of these ordinarily flexible muscles can give rise to such common health problems as high blood pressure (also called *hypertension*), heart attacks, and strokes.

Fatty substances, called *lipids,* are involved in producing atherosclerosis. Being fats, they do not mix with water, which is the basic medium of the body. To move around the body, lipids must combine with a specific class of proteins with which they form complexes, called *lipoproteins*. Other proteins, located on cell surfaces, serve as receptors that attach lipoproteins to various cells, where they carry out their functions.

In addition, numerous enzymes (all of which are proteins) are involved in the metabolism of lipids and lipoproteins. As the amino acid sequences of all these proteins are specified by DNA, numerous DNA sequences can be implicated in the development of these health problems.

Like many chronic health conditions, high blood pressure and related disorders often cluster in families. Most physicians therefore assume that "genetic risk factors" are involved, especially in families whose members encounter these problems at a relatively young age. In addition, everyone recognizes environmental or "life-style" risk factors, such as inappropriate diet, lack of exercise, and smoking.

By using the term life-style, I do not want to imply that these factors always are the result of free choice. Many economic, cultural, and other social factors, including corporate efforts to promote smoking and other clearly unhealthful habits, affect what people eat or drink, how much time and energy they can devote to healthful exercise, and whether they become addicted to cigarettes, alcohol, or other drugs that increase the risk of heart disease or strokes.

Like diabetes, the health problems related to atherosclerosis result from an interplay of genetic and environmental factors. Some, but not all, of the proteins involved in lipid transport and metabolism are known, but so far no one has identified the DNA sequences involved in their synthesis. At present, research at the molecular level relies on identifying RFLPs that can be correlated with the appearance of one or another of these conditions (see chapter 4).

We have seen that such correlation studies require an extensive data base, which catalogues variations within large families. It may be that researchers will eventually find genetic markers linked to specific conditions. So far, however, such studies have little if any predictive value for any specific individual, which is, after all, what people want.

Even if researchers manage to come up with tests that will tell some people that they have a higher than average chance of getting atherosclerosis, this will again raise the question of whether people benefit from knowing that they may be at special risk of developing a serious condition long before they experience symptoms. The argument in favor of providing that kind of information is that it may permit people to take special precautions, such as to eat less saturated fat and less red meat, to stop smoking, and so on. However, warning people of future illnesses is as likely to make them fatalistic as to spur them to action.

People have trouble keeping their good resolutions, whether or not they have hints of increased risks. Many personal and cultural factors regulate our "choices." It is hard to convert people who have grown up believing that butter, whole milk, and red meat are good for them to a diet of tofu, brown rice, and lentils, healthful though that might be. And no matter what the predictive tests may say, it is not surprising that people often fail to change their lives accordingly. If they can change, living in healthful ways is beneficial for everyone. Since risks of future disease cannot be predicted reliably, it would be better to help everyone to eat and live more healthfully, rather than singling out a few "high-risk" individuals.

Recently, some physicians have suggested screening the cholesterol levels of children in order to identify those who may be in danger of developing high blood pressure and heart disease as adults. Other physicians disagree.[11] As far as I know, no one is yet suggesting predictive prenatal tests, but I am sure someone will before long. None of these predictions makes any sense for conditions that are so variable in their age of onset and their extent, and that are affected by many nongenetic factors.

Before we leave this subject, I want to look at an issue that has come up repeatedly. High blood pressure is more prevalent among African Americans than among Euro-Americans, as are almost all chronic and acute medical conditions. Many economic and social reasons have been proposed to explain this disparity, all of them obviously related to racism. But, as always, some experts have managed to come up with genetic explanations.

After all, they say, African Americans are black and Euro-Americans, white, and that is clearly genetic. So, why not assume that there are genetic differences in other biological characteristics, such as susceptibility to various health conditions? This reasoning rests on a fallacy. "Black," in the United States, is more a political definition than a genetic one. Few, if any, African Americans are of pure African lineage. Rape and coercion of black women, as well as choice, have resulted in considerable genetic mixing. Since any hint of African descent is enough to stamp someone as black, the United States is full of "black" people who are genetically at least as European as they are African. Their "blackness" is a matter of arbitrary definition, not of biology. In fact, there are more genetic differences among Africans or among Europeans than there are between "blacks" and "whites" in the United States.[12]

Recently Thomas Wilson and Clarence Grim of the Hypertension Research Center at Charles R. Drew University in Los Angeles have offered a novel explanation for why African Americans are more susceptible to developing hypertension than either Euro-Americans or Africans are. They suggest that some African captives had alleles that made them conserve salt. This could have enabled them to survive the fierce heat as well as the high fevers, vomiting, and diarrhea they experienced on the slave ships and on plantations better than individuals who lacked such mutations. The researchers argue that these selective pressures have resulted in an inherited tendency to conserve salt, which is now a liability and leads to hypertension and associated disorders.[13]

I am skeptical about this explanation. To exert noticeable effects, selective evolutionary pressures must act over many generations. Even such cataclysmic events as the slave transports and slavery are not likely to have affected the distribution of genes in the population in this manner. Fatimah Jackson, a biological anthropologist, has argued that the historical events accompanying slavery—both the stresses and the mixing of previously separate populations—probably increased genetic diversity among African Americans rather than reducing it.[14] It seems likely that social and economic disadvantages and other effects of racism induce chronic levels of stress and that this, combined with the high salt content of the present-day American diet, predisposes African Americans to hypertension and the disorders associated with it.[15]

CANCER

Biologists are fascinated by cancer, not just because it is a serious health condition, but because it is a manifestation of growth gone awry. If scientists understood how cancers develop, they would understand a good deal more than they now do about the ways other tissues function and maintain themselves.

As we saw in chapter 4, every cell in my body arose from a single cell produced by the fusion of one of my mother's eggs with one of my father's sperm. Therefore, all my cells contain the same set of chromosomes and genes. And yet, muscle cells look and function differently from skin cells, brain cells, or the kinds of cells that circulate in my blood or lymph. All these cells contain the same DNA, but day in and day out each continues to behave the way it is supposed to behave. This is so despite the fact that all of them once had (and to a large extent still have) the genetic potential to behave like any of the others.

Ordinarily, the cells in a particular tissue divide on a schedule that is characteristic of that tissue. Even after I stopped growing, the *epithelial cells* that lie on the outer surface of my skin have continued to divide at a fair clip. And so have the epithelial cells that line my lungs, digestive tract, and the milk ducts of my breasts. By contrast, my nerve and muscle cells stopped dividing long before I stopped growing. The cells in my liver are somewhere in between, in that they have divided rarely since I stopped growing, but can divide in response to special demands.

If a mutation happens in a cell of a specialized tissue, that cell may stop behaving like a proper member of that tissue and begin to divide on its own schedule. This can start a growth that is not integrated into that tissue's normal metabolic pattern and function. The growth can become a nodule or a polyp and may be the beginning of a tumor. If the growth remains confined to the tissue in which it arises, it is usually considered benign. When a benign tumor gets too big or begins to interfere with that tissue's functions, it can be removed by cutting it out. (This is the case with the rather common basal cell cancers of the skin.)

In the kinds of cancers people worry about, not all the cells that form the growth stay in the tissue in which they originated. Some of them break loose and move into the circulating blood or lymph, where they can travel to distant tissues, take hold and produce further growths, called *metastases*. Metastases can make cancer deadly if they invade vital organs, such as the liver or brain, or when they become too numerous to be removed by surgery. The progression from the early to the final cancerous stages can take many years. In the normal life of a tissue, this process is probably initiated repeatedly, but is usually interrupted spontaneously so that no cancer develops. To control the effects of cancer, biologists need to learn a good deal more not just about how to keep the initial mutation from happening but also about how to keep the mutated cells from growing and proliferating.

Just as epithelial cells and muscle cells divide at very different rates, they also differ in the likelihood that they will develop cancers. In general, the greater the rate at which the cells in a tissue normally divide, the more readily a cancer can develop in that tissue. Presumably this is so because mistakes (mutations) are most likely to happen during cell division, when the DNA is duplicated. For one thing, the duplication process is not perfect. For another, it is at this time that DNA is most sensitive to carcinogens. Because epithelial cells keep on dividing throughout life, cancers arise relatively frequently in the skin, lungs, digestive tract, bladder, or in women's breasts.[16]

Where does genetics come into our understanding of the mechanisms of cancer? At this point most scientists agree that gene mutations have something to do with cancer, but the relevant mutations almost always occur in the differentiated cells of an individual. Ordinarily, when scientists speak of "cancer genes," they are not referring to the genes that parents transmit to their children in eggs or sperm. Since media reports about "cancer genes" often imply the opposite, it is important to get this straight. The scientists are speaking about genes that one cell passes to another as it divides within the tissues of the same individual—for example, when skin cells slough off and are replaced by new skin cells. When we hear the terms, "cancer genetics" or "cancer genes," we must understand that only rarely does a cancer mutation find its way into an egg or sperm and so get transmitted from a parent to his or her children. In fact, identical twins are not much more likely than other siblings to develop the same cancers, despite the fact that they were born with the same genetic material.[17]

There is general agreement that cancer-causing agents, or *carcinogens*, in the environment increase the likelihood that cancer-initiating mutations will happen. Environmental carcinogens, such as chemicals, radiation, and probably viruses, are responsible for between 70 and 90 percent of cancers.[18] The American Cancer Society reports that in the five years between 1984 and 1989 the death rate from cancer more than doubled in Alaska and increased by 50 percent or more in Nevada, Hawaii, and Puerto Rico. Clearly, this increase does not reflect a sudden change in patterns of inheritance, but is related to industrialization and urbanization, and in the case of Nevada to nuclear tests, with accompanying increases in environmental carcinogens.

Presumably, environmental factors play a part in the origin of cancers by increasing the probability of mutations. But no one believes that a single mutation is enough to produce a cancer. As scientists currently understand it, once a mutation has happened that can initiate a pattern of uncontrolled cell divisions, a number of further events must occur in that same cell before an actual cancerous growth develops and spreads beyond the original site. Factors in the environment are thought to elicit further mutations which can promote or restrain the subsequent conversion of "precancerous" changes to cancers.

Oncogenes and Anti-Oncogenes

At present, molecular biologists implicate two kinds of genes in the origin of cancers. They call these *oncogenes* and *anti-oncogenes* (*onco* is Greek for mass or tumor). Oncogenes are thought to be involved in the synthesis of proteins that promote cell division. Anti-oncogenes apparently are implicated in the synthesis of proteins that inhibit, or otherwise counteract, the effects of oncogenes.[19] Segments of DNA have been isolated that appear to fulfill these functions.

Using a cell line cultured in the laboratory from a human bladder cancer, scientists have isolated a gene that is closely related to an ordinary human gene, but different enough to initiate uncontrolled growth—the start of a cancer. A gene that can be made to mutate into an oncogene is sometimes called a *proto-oncogene,* but it is not clear whether only special kinds of genes can be transformed into oncogenes or whether any gene can be a proto-oncogene. Apparently a single base change can convert a gene into an oncogene. Also, it seems that if the same gene is transformed into an oncogene in different tissues, it gives rise to tumors that are characteristic of the tissue in which the mutation has taken place.

Besides base changes, other transformations can also convert genes into oncogenes. For example, the transposition of a gene from one chromosome to another can have this effect. Some oncogenes seem to have undergone no base change at all; instead, the "mutation" consists of the cell having many more copies of that gene than the usual two. This seems to make the cell divide too often and start a cancerous growth. Which proteins are associated with promoting or hindering such changes is still a mystery, but presumably their activities can distort the normal pattern of growth.

Anti-oncogenes, also called "tumor-suppressor genes," enter this picture by virtue of the fact that normal growth represents a balance between growth-promoting and growth-limiting metabolic activities. Anti-oncogenes are thought to specify the synthesis of proteins that restrain genes from serving as oncogenes, and so keep cancers from forming. For this reason, a carcinogenic mutation of an anti-oncogene is one that suppresses its usual metabolic effects. There is another difference in the ways oncogenes and anti-oncogenes are related to cancer growth. Whereas oncogenes appear to be generated by a mutation of a precursor-gene (a proto-oncogene) in the tissue in which the tumor appears, and therefore cannot be transmitted from parents to their children, there is some evidence that mutated anti-oncogenes can become part of the genetic line.

A mutated allele of a gene called p53, believed to be an anti-oncogene, appears to be associated with a wide variety of cancers, including colon, breast, and lung cancer. In family studies, people who had this allele developed proportionately more cancers than did relatives with the normal p53 allele.[20] Yet, it is important to understand that many of the people with the mutant allele did not develop cancer and, conversely, most people who develop cancer do not carry the mutated p53 gene.

As I have said, there is reason to believe that it takes more than one mutation to initiate the growth of a tumor. The production of an oncogene or a faulty anti-oncogene is only one in a series of necessary events. Molecular biologists speculate that different oncogenes may enter the chain at different stages but, here again, they tend to think in terms of linear sequences, which may not be the appropriate model at all. Since cancer represents a change in the normal pattern of growth and development, genes, proteins, other metabolites, and a variety of environmental factors no doubt are involved in a variety of complex and interrelated ways.

Cancer Prevention and Cancer Screening

If cancers result from several successive mutations or other events, this would explain why decades usually elapse between the initial change and the appearance of a cancerous lesion. Presumably that is why, by and large, cancer is a disease of older people. It may also explain why cancers that occur in children or young people tend to progress rapidly, because here the initial events happen at a young age, when cells in all the tissues are still dividing frequently.

Because the natural history of most cancers involves such a long lag time between the initial event and the appearance of symptoms, cancer biologists and physicians who specialize in cancer therapy (called oncologists) believe that if they could detect aberrant cell proliferation, they could stop or at least slow it before it becomes a cancer. Unfortunately, except for the cervix and skin, most epithelia (which, as we saw, are where cancers arise most frequently) are relatively inaccessible—in the lungs, breasts, digestive tract, pancreas, prostate gland, other glands, and so on—and this makes early detection difficult.

Since about 80 percent of cancers are of environmental origin, it should not surprise us that most cancers (56 percent) occur in the epithelia that communicate with the external environment: skin, colon, lungs, stomach, and cervix. Another 36 percent occur in the internal epthelia of the breasts, prostate, ovaries, bladder, and pancreas. Only 8 percent occur in bones and associated supporting tissues and in the blood-forming organs (these are the leukemias and lymphomas). Although physicians recognize some two hundred varieties of cancer, a dozen or so varieties account for about four-fifths of cancer deaths, and cancers of the lungs, colon, and breasts account for about half of these.[21]

It is important to realize that such statistics obscure as much as they reveal. The incidence of different forms of cancer has varied greatly over time and between different countries and different regions in the same country. The worldwide incidence of some common cancers can vary as much as three hundred-fold from one region to another. Lung cancer was rare in the nineteenth century, but now it is the commonest form of cancer in many countries, including the United States.

In the United States, the death rates from all forms of cancer are highest in the Northeast (New York, New Jersey, Connecticut, Massachusetts, Rhode Island) and lowest in Utah, Wyoming, and Idaho. As the molecular biologist John Cairns writes, "The implication is that about a third of the cancer deaths in the high-risk states would not have occurred if the victims had lived in the West."[22] The rates for individual forms of cancer vary even more. In the United States, poor people have higher death rates from cancers than rich people have (as is true for most diseases), but the incidence of some common cancers, such as cancer of the breast or prostate, is higher among the rich. Particularly in the case of prostate cancer, this apparent disparity may actually be a result of better diagnoses among wealthy men. Prostate cancer develops slowly, and usually manifests itself late in life. Undoubtedly, many men who do not get regular care have prostate cancers, but die of other causes without these cancers being detected.

Cigarettes and tobacco products account for about 30 percent of all cancers and lung cancer constitutes about one-quarter of cancers being diagnosed. (To put this figure in perspective: Tobacco is the largest non-food cash crop being grown in the world and the six major U.S. cigarette companies produce 600 billion cigarettes a year and direct their advertising primarily at young people and at poor people.) In addition to tobacco and industrial pollutants, diet and nutrition are thought to provoke up to 35 percent of all cancers, especially colon-rectal cancer.[23]

Breast Cancer

Let us stop briefly and look at what is known about breast cancer, since we tend to hear a great deal about it. In 1991, the American Cancer Society publicized the statistic that, in the United States, a woman's chance of getting breast cancer is one in nine. Since this number has been widely disseminated, it is worth spending a moment to discuss what it means.

Popular articles on breast cancer tend to emphasize the need for women to start watching for the signs of the disease in their thirties, so many young women think of the 1 in 9 figure as having an immediate relationship to their lives. However, breast cancer is by and large a disease of older women. The *New York Times* reports that "1 in 9 is the *cumulative* probability that any woman will develop breast cancer sometime between birth and age 110" (italics mine).[24] "Cumulative" means that this is the probability over a woman's whole life, not at any one time. In fact, the probability that a 35-year-old woman will get breast cancer by the time she is 55 is about 1 in 40, and the probability that she will die from it by 55 is only about 1 in 180.[25] Even for older women, the probabilities at any one time never get nearly as high as 1 in 9. How these numbers work may become clearer if we look at the pattern of breast cancer diagnosis in an average group of women. In table 1, we follow a

TABLE 1.

Incidence of Breast Cancer in a Hypothetical Group of One Hundred Women

Age	Incidence of Breast Cancer	Number of Women Alive	Probability of Developing Cancer in that Decade
30–40	1	100	1 in 100
40–50	1	100	1 in 100
50–60	2	100	1 in 50
60–70	2	90	1 in 45
70–80	2	70	1 in 35
80–90	2	50	1 in 25
90–100	1	30	1 in 30
100–110	0	2	indefinite
Total	11	100	1 in 9

NOTE: Over the entire eight decades, from age 30 to 110, eleven of the one hundred women—one in nine—will develop breast cancer. Thus, though the probability of developing this cancer in any single decade starts at one in a hundred and never exceeds one in twenty-five, for this group the cumulative probability of developing breast cancer over their lifetime is one in nine. (This is not a breakdown of real population figures.)

theoretical group of one hundred women with a cumulative risk of 1 in 9 of developing breast cancer, looking at the incidence of the disease by decades between the ages of 30 and 110.

Looking at this group in ten-year periods, we find that only one out of the hundred women gets breast cancer between the ages of thirty and forty, so we can say the chance of getting breast cancer in that decade is about one in one hundred. Between the ages of fifty and sixty, two get the disease, so the probability for that decade has climbed to one in fifty. At ages eighty to ninety, only two more women get breast cancer, but half of the women have died of this or other causes. Therefore, these two represent a one-in-twenty-five probability for women in that age range. Looking at the chart as a whole, we see that eleven of the original hundred women get breast cancer sometime in their lives, in keeping with the one-in-nine figure. However, at no time in her life is any one woman's probability of getting the disease within the next ten years higher than one in twenty-five.

It is important to understand these relationships because, when a doctor tells a woman that she has a one-in-nine chance of getting breast cancer, she may think that there is one chance in nine that a mammogram taken at that point will show her to have the disease. In fact, if she is forty, the probability that she will get breast cancer that year is only about one in one thousand. Even when she is sixty, the chance that a woman will have breast cancer in the next year is only about one in five hundred.[26]

What is more, all of these numbers are for the "average woman." The probability that any particular woman will get breast cancer depends on her so-called "risk factors." Most women who get breast cancer have no apparent risk factors, but a woman with all of the classic risk factors has a somewhat higher probability of developing breast cancer than does a woman with no risk factors. Risk factors include having a mother or sister who has had breast cancer, early onset of menstruation or late onset of menopause, not bearing a child before age thirty and, at least in some cases, high dietary fat or alcohol consumption.[27] However, these risk factors should not be taken at face value. As the *Times* article says, citing Dr. Patricia Kelly, director of medical genetics and cancer risk counseling at Salick Health Care in Berkeley, California, "Even having two close relatives die of breast cancer is not a death warrant. . . . If, for example, the relatives died in their 80's, a woman's risk would be no higher than normal." As for not having an early pregnancy, Kelly points out that "there are many reasons that women do not give birth, and not all increase risk."[28] Nancy Krieger, a cancer epidemiologist, has suggested that, before we accept the current list of risk factors, we need to learn a good deal more about the biology of the breast and its relationship to how women live: when they begin to menstruate; their sexual habits; whether and at what age they have an abortion or bear a first child; whether and how long they breastfeed; and so on.[29]

So, why is the American Cancer Society publicizing the one-in-nine figure? The *Times* quotes Joann Schellenback, a spokesperson for the society, as saying, "The 1-in-9 is meant to be a jolt. We use it to remind people that the problem hasn't gone away." Schellenback goes on to say that "many younger women look at nine of their friends and think 'One of us is going to get cancer this year.' The truth is that one of them *will* get cancer in her lifetime—but probably not until she's over 65."[30] Even worded this way, the example sounds scarier than necessary. It might be better to say that all nine women will die some time, and eight of them will die without ever having breast cancer. The ninth will develop breast cancer, but has two out of three chances of surviving the cancer and dying later of other causes. What is more, both Schellenback's example and mine are somewhat misleading, as it is quite possible that none of the nine women will get breast cancer. It is also possible that two of them will. The one-in-nine figure is not a law; it is only an average probability and therefore has little predictive power.

The American Cancer Society is using the one-in-nine figure to "jolt" women into getting regular breast examinations and mammograms. The problem is that such scare tactics may frighten women into undergoing mammograms too early in life or at too frequent intervals. As I said at the beginning of this chapter, this itself may increase the risk of developing breast cancer for some women.

There is another problem. A study of records of consultations for breast disorders at Ottawa Civic Hospital, published in the medical journal *Lancet* in 1989, found that 93 percent of mammograms that showed "signs/symptoms of breast cancer" proved to be false alarms.[31] I am not suggesting that mammograms are of no use, but women have to have more and better information than most of them are now getting. They need to be able to assess the risks they face at a particular time and make their decisions accordingly. Generalizations about "average women" and scare figures about "cumulative risks" are less than helpful.

Breast cancer is not life-threatening as long as it is confined to the breast. What makes this cancer dangerous is that it sometimes metastasizes, most usually to the lungs, liver, or bones. This is why detecting breast cancer before it spreads would make a big difference. Unfortunately, with current techniques this is not always possible. Breast cancer is not just one disease, but a family of diseases. Some forms of breast cancer spread even before a lesion can be seen on a mammogram. Others remain localized in the breast until well after a lump is large enough to be detected by self-examination or a routine medical checkup.

Since early detection does not work in all cases, it is tempting to look for "preventive therapies." The National Cancer Institute, which is one of the research institutes within the National Institutes of Health, has recently funded a prospective study in which eight thousand healthy volunteers, age thirty-five and over, who are thought to have special risk factors in their history will be given the drug *tamoxifen*. The incidence of breast cancer among these women will be compared with that found in a comparable group, who will be given dummy pills and will serve as "controls."[32]

Tamoxifen is an estrogen antagonist, and has been used for a number of years to treat women who have breast cancer, or to prevent a recurrence of the cancer after surgery. It is thought to act by blocking estrogen receptors in cancers whose growth is promoted by estrogen, and there is some evidence that it may also retard the growth of other forms of breast cancer.

Some women's health advocates support the tamoxifen study for its potential benefits, but others have voiced their opposition. In March 1992 Britain's Medical Research Council withdrew its support for a comparable study, pending further toxicological experiments.[33] Opponents of the study consider it too risky to give healthy women a powerful drug like tamoxifen, which can elicit blood clotting problems and is thought to increase the incidence of uterine and liver cancers and other liver diseases, as well as causing hot flashes, vaginal bleeding, menstrual irregularities, and possibly cataracts.

In the tamoxifen trial, researchers predict that, while the drug will prevent some sixty-two breast cancers and fifty-two heart attacks, it may cause thirty-eight uterine cancers and several deaths from blood clots in the lungs.[34] Arguing against this kind of experiment, Dr. Adriane Fugh-Berman, speaking for the National Women's Health Network, says that any program directed at a healthy "at-risk" population "should be extremely safe—preferably health promoting and . . . at least nontoxic. . . . We're afraid these tamoxifen intervention trials are really going to set a precedent for experiments in disease substitution—a concept we don't like."[35]

I strongly question the ethics of initiating this study, and am especially wary because, once again, a pharmaceutical firm is party to the debate. ICI Pharma, the manufacturer of tamoxifen, could make huge profits if this drug were prescribed not just to treat the subset of women who have breast cancer, but as a "preventive treatment" for healthy women thought to have certain "risk factors."

Cancer Prevention versus Cancer Therapy

Cancers develop slowly, probably over decades and, as we have seen, require a number of events. It makes sense that if the early stages could be detected and the activation and promotion of the cell into cancers counteracted, cancerous growth could be stopped. Indeed, as I have said, tiny cancers probably are initiated and quickly stopped repeatedly during everyone's life history, without our ever becoming aware of them. The cells in our immune system are thought regularly to mop up precancerous cells soon enough so that they do not become foci of further growth and develop into a cancer.

Early diagnosis could do a lot, but unfortunately it is so far usable mostly for easily accessible skin and cervical cancers. It is much more problematic for cancers of the deeper-lying epithelia, where the methods of diagnosis can themselves be invasive or dangerous. Therefore, we need to do a great deal more to prevent the initial events. This means decreasing the levels of radiation and environmental pollutants to which people are exposed. Speculations about genetic "predispositions" distract people from the need to make such environmental changes.

The frequently political nature of such geneticization becomes obvious when we look at the various research and news reports that attribute differences in health or in psychological or social characteristics between African Americans and Euro-Americans to inherent biological "tendencies." Only the rare scientist or reporter acknowledges that racism affects all aspects of growing up and living with dark skin, and that it is therefore fruitless to quantify and scientize this or that biological parameter in order to "explain" such differences.

Data show that the higher cancer mortality among African Americans cannot be attributed to their race, but rather is due to the fact that disproportionate numbers of African Americans are poor and therefore have inadequate education, diet, and access to medical care.[36] This has led some scientists to argue for better access to health care. But even more important is the chance to stay healthy. That requires profound changes in the distribution of incomes and wealth and in education and job structures.

Meanwhile molecular biologists continue to promote the idea that cancer can only be "solved" by learning more about the molecular mechanisms involved. When I. Bernard Weinstein, of the Institute of Cancer Research at Columbia University, addressed an annual meeting of the American Association for Cancer Research, where he received a special award, he said:

> Epidemiological studies provide evidence that environmental factors (external agents such as chemicals, radiation, and viruses) play a major role in the causation of the majority of human tumors. This is a highly optimistic message, since it implies that cancer is largely a preventable disease.

If you were to guess, what would you think he would say next? What he does say is:

> To meet this challenge we must, however, understand the mechanism of cancer causation at the cellular and molecular levels and, in a parallel effort, develop new laboratory methods that can be used to identify specific causative agents in humans. The approach must be comprehensive since it is likely that human cancers are due to complex interactions between multiple factors, including the combined actions of chemical and viral agents.[37]

I cannot help but wonder why Dr. Weinstein does not urge us to do everything possible to lower the exposure to the environmental carcinogens he mentions, by making the necessary economic, social, and political changes. Why do we need to wait for more scientific research? The answers are clearly political. Most of the self-destructive behaviors that lead to cancer are encouraged by corporations and governments, which subsidize, produce, advertise, and market products that they know to be dangerous. Lowering exposures would require broad-based changes. It is far easier and more convenient for scientists to pretend they will conquer cancer by studying the molecular transformations of genes and cells and coming up with diagnostic tests. Yet many health plans do not pay for the new screening tests and, of course, many people in this country are not covered by health insurance of any kind. As always, tests will be accessible primarily to the wealthier, hence healthier, segments of the population.

SEVEN

. . . .

"Inherited Tendencies": Behaviors

WHICH BEHAVIORS?

There is a big difference between associating genes with conditions that follow a Mendelian pattern of inheritance and using hypothetical genetic "tendencies" to explain complex conditions such as cancer or high blood pressure. Scientists make a further leap when they suggest that genetic research can help to explain human behaviors.

First of all, the criteria used to define behaviors, lump them into categories, and decide which of them are to be called normal and which pathological are based on a great many assumptions and decisions. It is true that anything people do involves their physiology, and hence their DNA. However, scientists only look for genetic components in behaviors which their society considers important and probably hereditary.

European peoples read from left to right, while Semitic peoples read from right to left, and many Asian peoples read from top to bottom. No one has suggested that these are inherent racial characteristics. However, traits like violence, dishonesty, and intelligence have often been considered hereditary and racially linked. These selections have been made for historical and cultural reasons and are in no way scientific, and yet they are the starting point for the discussion of genes and their place in behavior.

The behaviors we will look at in this chapter are connected only by the fact that they have been stigmatized by some segments of our society, and that scientists have recently claimed genetic links for each of them. Some of these links have been tenuous, others even less than that. However, because the claims have been made and widely reported, we need to examine the evidence.

93

HOMOSEXUALITY

Among the wide range of possible human behaviors, societies select some that they approve or extol, others they accept or tolerate, and yet others they dislike, condemn, and even persecute. These choices are part of the fabric that holds each society together.

All of us have a wide range of erotic feelings. Societies define some of these as sexual and regulate the degree and the ways in which we are permitted to develop and express them. Homosexual behaviors probably have existed in all societies, but our current perception of homosexuality has its roots in the late nineteenth century. That is when people began to consider certain sexual behaviors to be the identifying characteristic of those who practiced them. Homosexuality stopped being what people did and became who they were. As Michel Foucault writes in his *History of Sexuality,* until that time "the sodomite had been a temporary aberration; the homosexual was now a species."[1]

This way of categorizing people obscured the hitherto accepted fact that many people do not have sexual relations exclusively with one or the other sex. While they understood that the categories could sometimes blur, turn-of-the-century sex reformers such as Havelock Ellis and Edward Carpenter looked upon homosexuals as biologically different from heterosexuals. Homosexuals, or "inverts," were regarded as a separate category from heterosexuals who engaged in homosexual acts. As Jeffrey Weeks says in his history of homosexual politics in Britain, "Both 'inverts' and 'perverts' did the same things in bed . . . and the distinction relied on purely arbitrary judgements as to whether the homosexuality was inherent or acquired."[2] The reformers believed that "inverts" should not be punished for their acts, because their sexual orientation was biological and not a matter of choice.

Many modern researchers continue to believe that sexual preference is to some extent biologically determined. They base this belief on the fact that no single environmental explanation can account for the development of homosexuality. But this does not make sense. Human sexuality is complex and affected by many things. The failure to come up with a clear environmental explanation is not surprising, and does not mean that the answer lies in biology.

However, many people seem to believe that homosexuality would be more accepted if it were shown to be inborn. Randy Shilts, a gay journalist, has said that a biological explanation "would reduce being gay to something like being left-handed, which is in fact all that it is."[3] This argument is not very convincing. Until recently, left-handed people were forced to switch over, and punished if they continued to favor their "bad" hand. Grounding difference in biology does not stem bigotry. Quite the contrary. African Americans, Jews, people with disabilities, and also homosexuals have been persecuted for biological "flaws," and even exterminated to keep them from spreading biological "contamination."

Questions about the origins of homosexuality would be of little interest if it were not a stigmatized behavior. We do not ask comparable questions about "normal" sexual preferences, such as preferences for certain physical types or for specific sexual acts that are common among heterosexuals. Still, many gay people welcome biological explanations and, in recent years, much of the search for biological components in homosexuality has been carried out by gay researchers.

In 1991, two research studies were published that suggested a major biological component in male homosexuality. Both received a great deal of press coverage, including a cover story in *Newsweek*.[4] In their scientific papers, the authors of both studies were careful not to make claims of a genetic basis for their findings. However, in interviews they said they believed in such a basis, and strongly suggested that their studies provided evidence for its existence.[5] In at least some cases, they also extrapolated their results to include lesbians, although none had been included in the research.

In the first paper, Simon LeVay, a researcher at the Salk Institute for Biological Studies in San Diego, claims that an area of the hypothalamus (a region at the base of the brain) is smaller in homosexual men and heterosexual women than it is in heterosexual men, and that the smaller size is linked to a preference for males as sexual partners.[6] This linkage presumes a size similarity for the area among heterosexual men and lesbians, but LeVay was not able to test that part of his hypothesis.

This omission is only one of the defects in LeVay's work. All of the brain tissues he studied were obtained from cadavers, so there was no way to determine the range or extent of the men's sexual orientation. The study included only nineteen "homosexual" men (including one "bisexual"), sixteen "presumed heterosexual" men, and six "presumed heterosexual" women. All the homosexual men had died of AIDS, which may have affected their brain tissue. (LeVay added six heterosexual men who had died of AIDS to his study, to answer this criticism. This change reduced the difference between the "homosexual" and "heterosexual" men.[7]) Though, on average, the size of the hypothalamic nucleus LeVay considered significant was indeed smaller in the men he identified as homosexual, his published data show that the range of sizes of the individual samples was virtually the same as for the heterosexual men. That is, the area was larger in some of the homosexuals than in many of

TYSON, BOXING AND RAPE
By Joyce Carol Oates

Newsweek

Is This Child Gay?

Born or Bred: The Origins of Homosexuality

the heterosexual men, and smaller in some of the heterosexual men than in many of the homosexuals. This means that, though the groups showed some difference as groups, there was no way to tell anything about an individual's sexual orientation by looking at his hypothalamus.

The second study tried to determine the extent to which homosexuality is inherited, by looking at a group of homosexual men (including some bisexuals) and their brothers.[8] Michael Bailey and Richard Pillard, researchers at Northwestern University and the Boston University School of Medicine, looked at 56 pairs of identical twins, 54 pairs of fraternal twins, 142 nontwin brothers of the twins, and 57 pairs of adoptive brothers. They found that the rate of homosexuality for the adoptive and the nontwin biological brothers was about 10 percent, a rate often attributed to the general population. The rate for the fraternal twins was 22 percent, and for the identical twins it was 52 percent.

The fact that fraternal twins of gay men were roughly twice as likely to be gay as other biological brothers shows that environmental factors are involved, since fraternal twins are no more similar biologically than are other biological brothers. If being a fraternal twin exerts an environmental influence, it does not seem surprising that this should be even truer for identical twins, who the world thinks of as "the same" and treats accordingly, and who often share those feelings of sameness.

Another environmental factor, homophobia, may have had an even more drastic effect on the results of this study. Bailey and Pillard did not simply study a random sample of homosexuals. The gay and bisexual men "were recruited through advertisements placed in gay publications in several cities in the Midwest and Southwest." So all the respondents read gay periodicals and responded to ads asking them about their brothers. Though the ad asked gay men to "call regardless of the sexual orientation of [their] brother[s]," men with gay brothers might well have been more likely to participate than men with brothers who were straight, especially if the brothers were homophobic or if the gay men were not "out" to their families. Since many people believe that homosexuality is genetic, a straight man who has a homosexual identical twin might well feel that his own sexual orientation was "suspect," and might find the subject threatening. Conversely, identical twins who are both gay might find the subject interesting, and be eager to participate in a study.

Despite these flaws, and the fact that the authors acknowlege some of them in their paper, *Science News* quotes them as saying: "Our research shows that male sexual orientation is substantially genetic."[9] This is not because Bailey and Pillard, or LeVay, have any wish to mislead. Their research is painstaking, their methods are described in detail, and the authors are careful not to make extravagant claims in their papers. Their assessment of their work is only betrayed by their readiness to believe results that fit their preconceptions. Indeed, readiness may be too weak a word. *Newsweek* quotes LeVay as saying, "I felt if I didn't find any [difference in the hypothalamuses], I would give up a scientific career altogether."[10] Bailey and Pillard are less extreme. They are careful to address possible flaws and admit that their results are less than conclusive. And yet, their work and LeVay's reinforce each other. An article in *Science*, titled "Twin Study Links Genes to Homosexuality," quotes Bailey as saying: "Our working hypothesis is that these [homosexual] genes affect the part of the brain that he [LeVay] studied."[11]

More recently, Bailey and Pillard have reported that, using the same research methods, they have obtained similar results in a preliminary study of female twins and adoptive sisters of lesbians.[12] Also, researchers at the University of California at Los Angeles claim to have found yet another relationship between brain structures and sexual orientation.[13] In this latter case the researchers ascribe the differences to hormonal influences in the womb rather than to genes, but the study shares many of the same methodological problems as LeVay's work.

Given the publicity accorded to such studies, more research will undoubtedly be done on this subject. Molecular biologists are now soliciting participants from extended families with "at least three gay men or lesbians," hoping to find DNA sequences they can link to homosexuality.[14] In view of the complexities of doing accurate linkage studies and the necessarily small size of the samples, such studies are bound to come up with plenty of meaningless correlations, which will get reported as further evidence of genetic transmission of homosexuality.

ALCOHOLISM

Before we look at the question of a relationship between genes and alcoholism, we need to ask what "alcoholism" is. The term is, after all, extremely elastic. As an article in the *Harvard Medical School Mental Health Review* says, "As the social stigma of alcoholism becomes milder, more and more people are defined as alcoholics or become willing to define themselves that way."[15]

While writing about this subject, I will use the terms "alcoholism" and "alcoholic," but I do not wish to be taken as endorsing this kind of labeling. I use the words simply because they are less clumsy than circumlocutions like "people who regularly drink excessive amounts of alcoholic beverages," and I will evaluate their meanings in the course of this chapter.

Let us start by looking at some arguments for and against the proposition that alcoholism is a disease and has a biological basis. In contemporary parlance, "healthy" and "sick" have become synonyms for "good" and "bad." We speak of "healthy relationships" and "sick jokes." Similarly, we medicalize many social behaviors, such as sex, love, and work habits. No wonder we medicalize the consumption of alcohol or drugs, especially when they can be shown to give rise to physiological symptoms of addiction. Yet, despite the fact that nicotine is six to eight times as addictive as alcohol and elicits withdrawal symptoms, smoking has escaped the disease label. Smoking is also responsible for many more deaths than are drugs and alcohol combined. By current estimates, if one includes fires and accidents, one in four deaths in the United States can be attributed to the consequences of tobacco smoking.[16] Nonetheless, smoking is not nearly as stigmatized as excessive drinking. This is due at least in part to the power of the cigarette industry, though to be fair, excessive alcohol consumption can be a good deal more problematic for someone's family and friends than smoking is.

The belief that alcoholism not only is a disease but is incurable has been popularized by the adherents of Alcoholics Anonymous (AA). Just as people whose cancer has been controlled by surgery or other therapies are usually advised to consider themselves "in remission" rather than cured, so people who have stopped drinking excessive amounts of alcohol are told by AA that they are, and always will be, alcoholics. Even if they never drink another drop of alcohol, AA advises them to think of themselves as "recovering alcoholics."

Judging by AA's success as a self-help organization, many people who have drunk to excess find this a useful way to think about their drinking and believe that it helps them to stay sober. Nevertheless, the philosopher and educator Herbert Fingarette argues that the disease model of alcoholism is outdated and that AA's insistence that even a single drink puts you not just on a slope but on a well-greased chute ("once an alcoholic, always an alcoholic") limits the success AA could have if it were willing to let "alcoholics" become occasional, social drinkers.[17]

Whether or not alcoholism is a disease, we may still ask to what extent it involves genes. After all, behaviors that are not considered diseases, such as talent at music or math, are sometimes thought of as having genetic components.

Here we find that the evidence is mixed. Some researchers have suggested that alcoholism not only runs in families, which could be due to habit and custom, but that it is genetic. Other studies, including one conducted by researchers at the University of Michigan over a period of thirty years,[18] showed that "the children of heavy drinkers were no more likely than the children of abstainers or near-abstainers to become heavy drinkers."[19] Furthermore, "not only are children of alcoholics *not* doomed to be alcoholics themselves, but several studies have shown that children of alcoholics who *have* developed a drinking problem do better at moderating their drinking . . . than other problem drinkers." [20]

According to the *Boston Globe,* Archie Brodsky, coauthor of *The Truth About Addiction and Recovery,* "said that [the belief that heavy drinkers produce heavy drinkers] has deep roots that stem from the untested observations of clinicians who treat alcoholics, from flawed studies that seemed to support the idea, and the success of Alcoholics Anonymous in emphasizing the view of alcoholism as a disease." [21] The opposite view, that alcoholism is a disease that can be inherited, diagnosed definitively, and cured with medical interventions, is illustrated in an article in the *New England Journal of Medicine.* The writer, a physician, optimistically points out that with predictive genetic tests it should be possible to detect alcoholism even "before the patient or the patient's physician became aware of its presence. . . . The possibility that persons at risk for alcoholism could be identified before they began drinking holds the exciting promise of true primary prevention." [22] Apparently the author does not even ask how one would know that these people would have become alcoholics. He is so enthralled by the prospects of "predictive genetics" and "preventive cures" that he fails to notice the absurdities in his scenario. He also avoids asking a logical next question: How about people who are not "at risk"? Can they drink as much as they want without ever having to worry about becoming alcoholics?

Such predictions are not helpful, and can even be dangerous. Prophecies about behavior can become self-fulfilling. Psychologist Stanton Peele, a specialist in addictive behaviors, writes that "indoctrinating young people with the view that they are likely to become alcoholics *may take them there more quickly than any inherited reaction to alcohol would have.*" [23]

The most frequently cited study linking genes and alcohol found that the incidence of alcoholism among adopted sons who had an alcoholic biological parent was 3.6 times the incidence among adopted sons whose biological parents were not alcoholics. [24] Since the sons of alcoholics were more than three times as likely to be alcoholics, even if they were not raised by their biological parents, this finding has led some people to conclude that the effect must be genetic. However, as Fingarette points out, such studies do not "come anywhere near warranting the conclusion that there is a unique disease of alcoholism which is genetically determined. . . . At best the studies suggest that heredity is one factor, among many, that pertains in a minority of cases." [25]

In fact, the study in question does not even suggest that much. The problems with adoptive-child studies are legion, as are the problems with twin studies.[26] It has been shown, for instance, that variations in the age of adoption can completely change study results. The usual practice of attempting to match children with adopting parents who are similar to their biological parents has to be taken into account. Also, in the case of alcoholism, the child of an alcoholic parent may have considerable environmental exposure to alcohol while still in the womb.

Even if we accepted this study's result, only about 18 percent of the sons with an alcoholic biological parent became alcoholics, as opposed to 5 percent of the sons whose biological parents were not alcoholics. Though these results may suggest some biological component, we must remember that 82 percent of sons with an alcoholic biological parent did *not* become alcoholics. Even if we were to grant a genetic component in alcoholism, it clearly does not have a strong determining effect and could not be used to predict anything about a person's relationship to alcohol.

An adoption study conducted in 1977 of biological daughters of alcoholics showed that the incidence of alcoholism in these women was no higher than in a control group of women whose parents were not alcoholics.[27] A more recent twin study of both sons and daughters of alcoholic parents came out with the same result; there appears to be no link for the daughters, though the sons showed some correlation.[28]

If the sons of nonalcoholic parents were less prone to alcoholism than the sons of alcoholics, this still would not provide a very good prediction of who would have alcohol problems. Since there are many more nonalcoholic parents than alcoholic parents, most people who become alcoholics have nonalcoholic parents. So, in societal terms, it hardly matters whether the genetic scenario is true.

People drink for many reasons, from peer pressure to loneliness. Some drink more than others, again for many reasons, and some of these come to be defined as alcoholics. Yet, as Ernest P. Noble, a psychiatrist engaged in the search for an "alcoholism gene," so succinctly says, "The environment is a tremendously powerful agent in producing alcoholism. But genes are easier to study."[29] It seems that as long as some people believe that genes create a "predisposition" to alcoholism, molecular biologists will try to identify such genes.

One method of associating genes with alcoholism runs more or less as follows: When alcohol is ingested, it gets transformed in various ways. It can be converted into larger molecules, such as sugars, or broken down into carbon dioxide and water. Enzymes are involved in all these transformations, and the levels of these enzymes, as well as the extent to which the corresponding genes participate, may differ in different people. It is easy for scientists to pinpoint such genes. Then it simply becomes a question of correlating differences in the involvement of one or another enzyme or gene with the extent to which different people get drunk from the same amount of alcohol. However, to construct a convincing connection between these genes and alcoholism, it would be necessary to show that the ease with which people get drunk runs parallel with the tendency to consume excessive amounts of alcohol, which is extremely problematic.

At present, rather than looking for genes involved in the metabolism of alcohol, molecular biologists are focusing on the gene on chromosome 11 that has been implicated in the synthesis of the dopamine receptor. This is the gene I mentioned in chapter 5, which is supposed to affect the ways in which brain cells communicate. In a recent issue, the *Journal of the American Medical Association* published two articles on this subject side by side. One claimed an association between this gene and alcoholism,[30] while the other denied the existence of such an association.[31]

Even the paper that claims the association does not suggest that the gene "causes" alcoholism. It merely suggests that this gene modifies the activities of other genes, which according to the article may contribute not only to alcoholism but to such behavioral disorders as schizophrenia, Tourette syndrome, and drug addiction. The gene is thought to effect these changes by virtue of its relationship to the dopamine receptor, which previously has been implicated in these and other behavioral disturbances.

An editorial in the same issue of the *Journal* throws its weight behind the claimed positive relationship of this gene to alcoholism, though the title emphasizes that the gene and alcoholism are "associated but not linked."[32] This should give us pause. The editorial makes such a distinction because the first article states that a particular DNA sequence on chromosome 11 is more common in the study's sample of alcoholics than it is in the general population. This is what the editorial calls an "association." But the article goes on to say that this sequence is not found in most alcoholics and is not linked with alcoholism when comparing individuals within family pedigrees.

In other words, the researchers looked at a small sample of alcoholics and found an RLFP pattern they had in common. However, when they compared family members who exhibited that pattern with those who did not, they found no link to alcoholism. This result should be a sufficient reason to consider the original observed association to be a coincidence, like when several people in a room have the same first name or the same birthday. Yet, the editorial claims that the study proves the DNA sequence is "likely to modify the expression of alcoholism rather than be a necessary or sufficient cause." This is grasping at straws.

If every correlation between a specific DNA sequence and the occurrence of a particular trait in a few people is taken to be meaningful, the scientific literature will soon be full of such claims. There will be enormous confusion if physicians and molecular biologists, in their desire to discover instances of genetic mediation, start searching the thesaurus for terms like "linked" and "associated" to fill the space between "cause" and "coincidence."

To say it again: There must be innumerable correlations between specific DNA sequences and traits that occur in some people. The task is to establish which correlations have functional significance.

PROBLEMS OF LINKING GENES TO BEHAVIOR

There is another angle that is being pursued in the search for genes "for" behavioral problems. Neuroscientists have described various receptors on the surfaces of cells that can combine with chemicals known to mediate "highs" or "lows." All such receptors are proteins, and can therefore be linked to genes. Some of the chemicals have been implicated in the psychological and physiological manifestations of alcoholism and other addictions as well as in different forms of mental illness.

Genes "for" schizophrenia or manic-depressive disorder are as questionable as genes "for" alcoholism. Despite the news stories we have read about genes for schizophrenia, we must understand that if one of a pair of identical twins, who of course have identical genes, develops schizophrenia, the likelihood that the other will develop it is about 30 percent.[33] Another way of saying the same thing is that the second twin has a better than two-in-three chance of *not* developing schizophrenia. The likelihood that an ordinary, first-degree sibling of a person with schizophrenia will also develop schizophrenia is about one in twelve.[34]

The behavioral psychologists Robert Plomin and Denise Daniels summarize a variety of observations by saying that the "environmental influences that affect psychological development . . . make children in the same family as different from one another as are children in different families."[35] Though this may be a bit extreme, it suggests how difficult, if not impossible, it will be to predict behavioral disorders on the basis of familial or genetic information.

It is possible to identify family and ethnic clusters, groups of genetically related people who manifest the relevant "symptoms" or "tendencies" to develop detrimental conditions or behaviors. But, in all these situations, there are so many environmental influences and such an intricate, ongoing interplay between environmental and inherited factors that the root of the condition becomes hopelessly difficult to find.

A recent study by a group of physicians in London suggests that people whose mothers had influenza during the second trimester of pregnancy may have an increased risk of developing schizophrenia. Thus, though these people were born with a "predisposition" to the condition, its cause is environmental.[36] A study of identical twin pairs in which only one twin in each pair had schizophrenia has found "subtle abnormalities of cerebral anatomy" in the schizophrenic twin. Whether the abnormalities caused the condition or vice versa, they cannot be genetic, since they are not shared by the genetically identical twin.[37]

With all this confusion, the presence of a genetic marker for a behavioral condition would not be particularly useful, even if it were found. Though the disease model or the genetic model can be helpful to some affected people, as with alcoholism, correlations between the conditions and specific base sequences of DNA do not add any useful information. Such correlations can neither predict the behavior for specific individuals nor yield treatments. Identifying a "culprit" DNA sequence is just a fancier way of saying that the condition runs in the family.

Since most diseases and many behaviors have biological correlates, there is never going to be any trouble identifying proteins that can be correlated with "symptoms." At the outset of any study, it will be unclear whether an observable correlation has sufficient biological significance to be worth pursuing or whether it has any predictive value. To answer such questions always requires looking at a wide range of people to test to what extent any correlation is meaningful.

What is more, the very effort to establish correlations raises serious scientific and ethical issues. How can predictions have scientific validity when a behavior may be altered by the very fact that it is being studied? And how can a responsible scientist ask people to participate in experiments when their participation may affect their behavior in ways neither they nor the scientist can foretell?

Pitfalls of Behavioral Research: The XYY Fallacy

There is another, basic, scientific issue. In order to set up correlations between a given genotype and a specific trait, it is wrong just to look at the people who manifest the trait and ask to what extent they share a specific genetic configuration. Rather, one needs to look at a random sample of the general population, in order to find out whether, or to what extent, people who exhibit that genetic configuration also manifest the trait. It is a common fallacy in correlational research to ignore this fact.

The best-known example of this mistake was the highly publicized claim of a link between the XYY chromosome anomaly and criminal behavior. From the early days of the eugenics movement, scientists have tried to find a genetic basis for criminality. These attempts have been spurred on by studies such as one recently reported in the *New York Times*, which produced the unsurprising statistics that "more than half of all juvenile delinquents imprisoned in state institutions and more than a third of adult criminals . . . have immediate family members who have also been incarcerated."[38]

Before we get into the biological discussion, let us remember that "criminality" is a social construct. Whether a given behavior is considered criminal depends on the context. Killing can be heroism or murder; taking someone's property can be confiscation or theft; until very recently, and still in many societies, rape has been considered normal sexual behavior if it happened within marriage. And, whatever the definition, incarceration is not a measure of criminality. People are incarcerated not because they have committed crimes but because they have been caught and have not been able to mount an adequate defense.

The XYY hypothesis was one recent attempt to find biological correlates for "criminal behavior." To understand it, we must first look at what the XYY condition is. As we saw in chapter 4, women have two X-chromosomes and are therefore said to be XX. Men have one X-chromosome and one Y-chromosome, and are said to be XY. Occasionally, when a sperm-forming cell goes through its reduction divisions, a mistake happens and a sperm ends up with two Y-chromosomes instead of just one. If such a sperm fuses with an egg, the embryo ends up having 47 chromosomes instead of 46, because it has three sex chromosomes. It is XYY instead of XY. All three sex chromosomes are duplicated at every cell division, so that the child ends up with the extra Y-chromosome in the nucleus of each of his cells.

In many cultures, being male is associated with being aggressive. Let us be clear at the outset that although some psychologists claim that there is an inherent, universal connection between maleness and aggression, this assumption has not been borne out by comparative research on different cultures. In our culture, a metaphoric connection has been drawn between aggressivity and testosterone, the "male hormone" that is secreted by the testes. Since the development of testes depends on the presence of a Y-chromosome, a link has been assumed between aggressivity and Y-chromosomes. (For a more detailed refutation of such claims, see my *Politics of Women's Biology*.[39])

In the early 1960s, several articles on XYY males appeared in the scientific literature. They posited that the presence of the XYY genotype was responsible for excessively aggressive behavior and, further, for what was loosely termed "criminality." In the most thorough study, Patricia Jacobs and her colleagues surveyed the distribution of the XYY genotype among a group of men incarcerated in a high security mental hospital in Scotland. They reported that this prison population contained about twenty times the proportion of XYY men expected for the general population. These men were also said to be unusually tall and "mentally sub-normal."[40] As a result, the authors attributed the mental illness, mental retardation, tallness, and aggressivity to the XYY genotype.

The Jacobs study stimulated a flurry of similar studies in mental hospitals and prisons in different countries, accompanied by headlines in the press about "criminal genes." When the dust settled, it turned out that there was no basis for the XYY hypothesis. Most incarcerated XYY men had not committed crimes of violence. More importantly, studies conducted in the population at large showed that all but a small percentage of XYY males lead ordinary lives and are not unusually aggressive. So, the assumption underlying this research was wrong.

However, before the XYY bubble burst, two scientists initiated a prospective study in which they proposed to screen all boys born at the Boston Lying-In Hospital in order to identify those who were XYY. They planned to observe the behavior of these boys systematically for many years both at home and in school, and to note any "abnormalities." The researchers also planned to provide the families with counseling, so that these families would be able to know what to do if their sons developed exceptionally aggressive behaviors.

This research program was challenged in the early 1970s by a group of progressive scientists associated with the Boston-based organization *Science for the People,* and it was eventually stopped.[41] *Science for the People* objected that, since the parents initially were not told about the purpose of the study or about its possible outcome, they were not in a position to give informed consent. The group further pointed out that, if the parents had been told everything they should have known, the study could not have been done properly. Once parents knew their son was XYY and that this might be linked to aggression, they would never know whether Johnny pushed Max because he woke up cross or because he was innately aggressive. If they went into a panic every time Johnny exhibited "aggressive" behavior, their anxiety would surely affect Johnny. (The Stanford psychologist Eleanor Maccoby has pointed out that "aggression" in children often is in the eye of the beholder: When Johnny pushes Max, if Max smiles, they are playing, but if Max starts to cry, Johnny has been aggressive.)

The Boston study baldly raises the issue of whether it is ever appropriate to do predictive experiments on random populations of "normal" individuals, since the very fact that the experiment is being done is likely to change these people's lives and so affect the outcome. This problem bedevils every attempt at doing behavioral genetics, whether it involves looking for genetic correlations with alcoholism and other addictive behaviors or with different forms of "mental illness." In view of the scientific, practical, and ethical problems of doing this kind of research, it is unlikely that the new techniques of molecular genetics will extricate this field from the swamp of contradictory findings in which it is mired. But, since these issues are of interest to our society, scientists will no doubt go on trying to find answers to such questions and the media will continue to publicize the claims and counterclaims. Given the confusion that inevitably surrounds any attempt to find causes for behaviors, it is understandable that people tend to accept whichever "facts" confirm their beliefs and preconceptions.

EIGHT

. . . .

MANIPULATING OUR GENES

CONVENTIONAL TREATMENTS FOR INHERITED CONDITIONS

Most current treatments for inherited conditions were developed before scientists knew how to determine the specific molecular composition of the DNA sequences implicated in such diseases. The treatments simply focused on alleviating symptoms. The most successful example of this is the treatment for phenylketonuria (PKU). Since PKU can serve as an example for a group of genetic diseases, sometimes called "inborn errors of metabolism," which includes galactosemia and lactose intolerance, it is worth looking at what this treatment involves.

PKU results from a person's inability to metabolize the amino acid *phenylalanine*, which is a component of many proteins. Most people have an enzyme in their tissues that converts phenylalanine to another amino acid, called *tyrosine*. In people who have PKU, this enzyme does not function properly. As a result, phenylalanine accumulates in the body, which can cause damage to brain cells and other tissues.

To prevent this damage from occurring, infants and children with PKU must severely limit their intake of proteins and receive supplements of tyrosine and the other amino acids essential for normal metabolism and growth. The children must adhere to this diet until they stop growing, and from then on can eat an ordinary diet.

Because this dietary treatment enables children with PKU to become healthy adults, in recent years an unanticipated situation has arisen. When girls born with PKU grow up on the modified diet and themselves begin to have children, it turns out that, though the elevated level of phenylalanine in their bloodstreams no longer poses a hazard to them, it can damage the fetus they carry. Therefore girls and women who have PKU are advised to remain on the diet until they are finished with childbearing.

108

Unfortunately, equally effective treatments do not exist for other common inherited conditions, such as sickle-cell anemia or cystic fibrosis. Nonetheless, treatments have been developed that make the symptoms less onerous. With sickle-cell anemia, some of the newer therapies have decreased the extent of the painful "sickling crises." And the frequent infections to which people with sickle-cell anemia are prone can be prevented or treated with antibiotics. For people with cystic fibrosis, daily physical therapy has been found to clear the mucus that threatens to block their bronchial passages and new drugs have come on the market that thin the mucus and make it more manageable. These measures enable them to breathe more freely, and reduce the incidence of respiratory infections, while antibiotics have helped keep infections under control.

In the case of some inherited conditions, such as congenital malfunctions of the thyroid, pituitary or adrenal glands, it has become possible to supplement the substances that are being produced in insufficient quantities. People with hemophilia, which is also inherited, can be treated by regularly supplying them with the blood clotting factor which they lack.

All these therapies were developed before scientists could pinpoint the specific genes implicated in these conditions. Where it is possible, treating symptoms has the advantage that the treatments can be increased, decreased, or stopped as needed. This raises no special medical or ethical issues beyond the usual ones that are involved in all medical interventions: access, equity, and the need for good communication between professionals and their clients. However, now that it is becoming possible to manipulate and alter genes, many scientists would like to get beyond treating the symptoms and change the associated segments of DNA.

MODIFYING DNA: SOMATIC CELL MANIPULATIONS

Ever since scientists realized that some conditions were associated with mutations in specific genes, they have argued that such conditions might best be treated by correcting the mutations themselves. To take a conceptually simple example: If a child has PKU, instead of modifying the diet to eliminate the symptoms physicians could "repair" the allele, replacing the mutated DNA sequence with a "normal" one.

In the scenario that is usually presented, physicians would put copies of normally functioning DNA sequences into the tissues in which the metabolic defect is most noticeable. To be effective—to "work"—the DNA sequence would have to be inserted into the chromosomes in the nuclei of cells that ordinarily produce the substance that is lacking or not functioning as it should. For example, if someone has sickle-cell anemia, a physician would try to introduce the DNA sequences involved with the synthesis of normal hemoglobin into cells in the bone marrow, where red blood cells are produced. If the procedure worked, these cells would then begin to synthesize normal hemoglobin. Scientists call such repairs at the tissue level *somatic gene therapy*.

I do not like the term "gene therapy" and shall avoid using it. It was introduced for its public relations value, before any gene manipulations were attempted, to suggest that such manipulations would be beneficial. To call gene modifications "therapies" assumes, without proof, that they will have therapeutic value. The term "therapy" implies a benefit to health, and that is why it gets used rather than the more neutral word "treatment," or a frightening word like "surgery." We should not accept the promise of benefits from gene manipulations before they have been shown to improve people's health.

Medical scientists have been preparing to do somatic gene modifications in humans for a number of years and have performed comparable DNA insertions in animals. However, before any such procedure can be used in humans, we need to be sure that the DNA is always inserted in the right place. Genes transposed into the wrong place may not function at all, or may lead to serious malfunctions, including cancers. Also, if the inserted piece of DNA happened to land in the middle of another gene, it could interfere with that gene's functions. Unlike many conventional treatments, gene insertions would be difficult, if not impossible, to reverse if the inserted DNA should prove to be harmful.

In recent experiments, conducted with isolated cells or in animals, scientists have managed to knock out specific DNA sequences on a chromosome and to insert others in the same location. However, this does not necessarily mean that the new sequences will function and be regulated properly and that they will not do more harm than good.

Another potential problem is posed by the *vectors* physicians use to carry pieces of DNA into the nuclei of the appropriate cells. In experiments on isolated tissues grown in tissue culture, inactivated viruses are often used to carry pieces of DNA into cell nuclei. Before such vectors are used in human beings, it is essential to make sure that they do not manifest unexpected and unwanted biological activities of their own.

Because researchers had to think through a wide range of precautions and test the procedures on animals, it took until the summer of 1990 for them to run a first experiment with human subjects. In this experiment a "foreign gene" was inserted into lymph cells of people who have malignant melanoma, a deadly form of skin cancer.[1] The DNA was not supposed to replace a faulty gene, but rather to increase these people's ability to produce antibodies that might help them resist the cancer.

The second gene modification experiment was performed in September 1990 on a four-year-old girl. This procedure is intended to help children who have inherited an extremely rare and life-threatening immunodeficiency disease similar to the acquired immunodeficiency syndrome (AIDS). Some children who manifest this disease appear to lack an enzyme, called *adenosine deaminase (ADA)*, which ordinarily participates in making antibodies. In this experiment, lymphocytes (a kind of white blood cell) were withdrawn from the girl and the gene that specifies the amino acid sequence of ADA was inserted into their nuclei. The modified cells were then transfused back into her bloodstream.[2] It looks as though the experiment was successful, and the girl has been able to produce sufficient amounts of the ADA enzyme to relieve her immune deficiency. Now, two more ADA-deficient girls are being treated similarly, receiving transfusions of their own modified cells every few months.[3]

Despite its success, several objections have been raised to this experiment. One is that a more conventional drug, which enhances the ability of children with this form of immunodeficiency to fight off infections, had been approved by the Federal Drug Administration. DNA manipulation therefore did not constitute the only available treatment for this condition, although it is a stated precondition for the approval of any gene modification experiment that there be no available alternative treatment.

A second objection is that one of the first experiments of this sort was undertaken with a child, who by definition cannot give consent. A third is that ADA deficiency, though life-threatening, is one of the rarest diseases known. Only some seventy children in the whole world are known to have it and only fifteen or twenty of these appear to lack adequately functioning levels of ADA, which is all this particular intervention can repair. So, although this manipulation may indeed benefit these few extremely sick children, it is quite proper to ask why so much time and so many resources are being devoted to treat this particular condition, when many children (and adults) the world over are suffering and dying from prevalent diseases for which treatments already exist.

Since these initial experiments were begun, many others have been proposed and some are being tried out. The experiments that could affect the greatest number of people involve the genetic mutation implicated in cystic fibrosis and a new way to tackle melanoma and other cancers.

People with cystic fibrosis lack a protein involved in transporting chloride ions across the membranes of epithelial cells. Scientists at the

National Institutes of Health have inserted the allele that specifies the composition of this protein, obtained from human chromosome 7, into the air passages of cotton rats. The rats shortly began to produce the human protein. Though in rats the gene was inserted directly into the trachea by surgery, the hope is that people could use less drastic methods, such as inhalers. The researchers caution that "the safety and effectiveness [of this procedure] remain to be demonstrated."[4] Ronald Crystal, a member of the research team, has raised the concern that the viral vector being used to carry the gene into the cells normally causes respiratory ailments. Although the team is using a modified form of this virus that cannot reproduce itself, the virus might revert to its normal state if it interacted with natural viruses in people's lungs. Another problem Crystal mentions is that people might develop immunity to the viral vector, so that the treatment, which would have to be administered periodically to be effective, would stop working.[5]

The newly proposed treatment for melanoma involves injecting trillions of copies of the gene that specifies an antigen protein directly into a tumor. The hope is that this protein will stimulate the person's immune system to produce cells that will attack the tumor. However, according to the New York Times's Natalie Angier, "Researchers warned . . . that it would be several years before the effectiveness of the approach for thwarting melanoma or any other disorder could be demonstrated." Yet Angier lauds the ingenuity of the researchers who, "limited only by the breadth of their creativity . . . are striving to devise gene therapies to better treat inherited diseases like hemophilia, cystic fibrosis, dwarfism and immune deficiencies, as well as chronic adult ailments like cancer and heart disease."[6] So, although we will not know for a long time whether and to what extent these sorts of treatments will be successful, the Times is already hailing them as the wave of the future.

Without question, during the next few years various DNA sequences will be inserted into cells and tissues in the hope of alleviating or curing diseases. Judging from past experience with organ transplants and heart machines, there will be plenty of volunteers eager to try these new treatments. But since the treatments are all experimental, it will take time before we will know whether each can relieve the health condition it is designed to remedy and what unexpected problems can arise.

In summary, such genetic interventions raise the same kinds of concerns as many other new procedures: They may involve unanticipated hazards and may not produce the expected results. Also, they can only address the health problems of relatively few individuals. To these people, the treatments may be a real boon, but unfortunately high-tech experiments drain resources away from the kinds of public health and medical measures that could improve the health of much larger numbers of people.

"Germ-Line Gene Therapy": Changing Future Generations

Attempts to modify the DNA in our reproductive cells (sperm and eggs) or in the cells of early embryos raise quite different and more troubling issues. If the DNA of these cells is altered, this will not just affect a specific individual, as in the previous scenarios. The altered DNA will be passed to future generations in the *germ line,* which is why this type of DNA manipulation is conventionally called *germ-line gene therapy.* Let me make clear how it differs from the sorts of gene manipulations that we have already considered.

If the DNA in a differentiated tissue, such as liver or skin, is altered by inserting or modifying all or part of a functional DNA sequence, this will only affect the person whose tissue is being modified. So, if it does no good or, worse yet, does harm, only that person will suffer the consequences. However, if one modifies the DNA in a sperm or egg or in the cells of an early embryo, the altered DNA will be copied each time these cells divide and will become part of all the cells in that future person, including her or his eggs or sperm.

This sort of manipulation is not intended to treat the health problems of actual people, but rather to alter the genetic makeup of hypothetical future people. Current attempts at germ-line gene manipulation involve the use of very early embryos, produced in a dish by in vitro fertilization. When the fertilized egg has divided into six or eight cells, the scientists remove one or two cells and test them to see whether they have the mutation the scientists are trying to remedy. Because at this stage all of the cells are still equivalent, removing a couple of them does not damage the embryo. If the embryo has the suspected mutation, the scientists could try to correct it through gene manipulation, before inserting it into the womb of the woman who will carry the pregnancy.

This is interesting science, but it is hard to see whom it would help. In the in vitro procedure, at least a half-dozen eggs are fertilized, of which only a few are usually implanted. The rest can be frozen and stored for later use, or discarded. Since there is a choice of which embryos to implant, one could select those which do not have the mutation; there is no reason to choose those which do have the mutation and then attempt to correct it. The one situation in which such embryo selection would not work is if all of the embryos had the mutation. This will only happen in the rare situations in which both parents have two copies of the same recessive allele, as for example if they both have sickle-cell anemia or the same form of cystic fibrosis. If such couples do not want a child with the same condition, there is a range of available options, from adoption to using a sperm donor. It hardly seems reasonable to develop germ-line gene manipulation for this purpose.

In terms of curing or treating genetic conditions, germ-line manipulations are completely irrelevant. There are no sick people who will benefit, and there are other ways to avoid passing on specific genetic traits. What is more, this technology could have some frightening consequences. As we have seen, if scientists alter the DNA of an early embryo those alterations will not only be incorporated into the cells of the person into whom that embryo may develop, but into the cells of her or his children, becoming a permanent part of the hereditary line.

This permanence raises troubling questions. It is not unusual for medical therapies to have unanticipated, undesirable effects. These are commonly called "side effects," though they may have more serious consequences than the intended effects. If a treatment produces an actual disease condition, that disease is called *iatrogenic,* which is Greek for "medically generated."

Iatrogenic conditions can be serious, even deadly, but often if a medication evokes untoward symptoms it can be stopped and, with luck, the symptoms will disappear. If ill effects show up in the next generation, as with DES or thalidomide, this is still not a permanent genetic change. However, if a genetic manipulation of the germ line turns out to be iatrogenic, medical practitioners will have become sorcerer's apprentices. The condition they have introduced will be beyond their control and it will be heritable.

By saying this, I do not want to overstate the potential significance I am prepared to assign to genes. I have said repeatedly that genes may turn out to play a considerably less important metabolic role than molecular biologists and geneticists tend to think they do. But the range and variety of the effects that may result from inserting or modifying chunks of DNA are unpredictable in any specific instance.

We must remember that a single gene may function differently in different tissues. The fact that scientists have linked a DNA sequence with a specific trait does not mean that the sequence has no other functions. It may participate in other metabolic reactions that scientists know nothing about. Tampering with DNA will have unexpected effects, and there is every reason to believe that some of them will be undesirable.

114

To introduce changes into an individual's hereditary line goes way beyond what we ordinarily think of as a justifiable medical intervention. Yet, if the DNA of early embryos proves to be easier to manipulate than DNA in the differentiated tissues of children or adults, some scientists will advocate manipulating the DNA in germ cells. If attempts at somatic gene manipulation have been less successful than promised, germ-line manipulations can be touted as a more effective way to get the same results, not for people with the conditions, but for their future children. However, if somatic gene manipulations are successful in some cases, people may be persuaded that germ-line manipulations are a logical next step.

As Edward Berger, a biologist, and Bernard Gert, a philosopher, point out:

> Past experience has shown that exciting new technology, including medical technology, generates pressures for its use. Thus, it is quite likely that if germ-line gene therapy were allowed, it would be used inappropriately. . . . In the real world researchers will overestimate their knowledge of the risks involved and hence will be tempted to perform germ-line gene therapy when it is not justified.[7]

Lest this sound unduly alarmist, here is a quotation from Daniel Koshland, a molecular biologist and editor-in-chief of *Science* magazine. Writing on the ethical questions posed by germ-line gene manipulations, Koshland muses about the possibility "that in the future genetic therapy will help with certain types of IQ deficiencies." He asks, "If a child destined to have a permanently low IQ could be cured by replacing a gene, would anyone really argue against that?" (Note the use of the word "cured" for averting the "destiny" of a "child" who would, at the time of the cure, be a half dozen cells in a petri dish.) While voicing some misgivings, Koshland continues:

It is a short step from that decision to improving a normal IQ. Is there an argument against making superior individuals? Not superior morally, and not superior philosophically, just superior in certain skills: better at computers, better as musicians, better physically. As society gets more complex, perhaps it must select for individuals more capable of coping with its complex problems. . . .[8]

Clearly, the eugenic implications of this technology are enormous. It brings us into a Brave New World in which scientists, or other self-appointed arbiters of human excellence, would be able to decide which are "bad" genes and when to replace them with "good" ones. Furthermore, the question of whether to identify the functions of particular genes or to tamper with them will not be decided only—or perhaps even primarily—on scientific or ethical grounds, but also for political and economic reasons. We need to pay attention to the experiments that will be proposed for germ-line genetic manipulations, and to oppose the rationales that will be put forward to advance their implementation, wherever and whenever they are discussed.

· ·

NINE

GENES FOR SALE

FUNDS FOR RESEARCH, PROFITS FOR BIOTECHNOLOGY

There are reasons why molecular biologists place such emphasis on genes, referring to them as "blueprints" of the organism. The belief that genes determine, and therefore can be used to predict, a wide range of significant traits and diseases is essential in order to marshall the popular and congressional support these scientists need. If genes can be implicated only in relatively rare conditions, such as Tay-Sachs disease, sickle-cell anemia, or cystic fibrosis, it is difficult to justify spending increasing amounts of money for the analysis of DNA at a time of shrinking budgets for other lines of biomedical research and for all sorts of social and medical services.

If, however, every health problem to which we may be susceptible, as well as the ways we grow, learn, and age are encoded in our genes, then every citizen and therefore every member of Congress will want scientists to find out everything they can about the human genome. If scientists convince us that we all have "genetic tendencies" to develop health or social problems, we all become candidates for genetic diagnosis and therapy. The benefits to medical research and development are clear, even if the benefits to us "patients" are questionable.

I do not mean to suggest that molecular biologists are deliberately deceiving people when they advertise the potential effectiveness of their work. Some may be doing that, but the more significant point is that they are members of this culture, which is ready to devote huge sums and much effort to eliminating biological causes of illness and death while at the same time accepting as inevitable a steady increase in the death toll from social causes. Such social causes include accidents and

117

violence, preventable environmental toxins, malnutrition, and infectious diseases, all of which injure or kill large numbers of people.

The pharmaceutical and biotechnical industries have invested a great deal of money in developing and marketing new products of genetic technology. Philip Abelson, former editor-in-chief of *Science* magazine and no opponent of biotechnology, writes that the major U.S. biotechnology companies spend 24 percent of their income on marketing and their salespeople make thirty million visits a year to doctors' offices to market their products.[1] But so far profits have been disappointing.[2] Products such as human insulin, growth hormone, and interferon are being produced, but the need for these substances is limited. These products are not going to turn the biotechnology industry into the equivalent of the computer industry of two decades ago and produce a new "Silicon Valley" or "Massachusetts miracle." For that, biomedical and biotechnical entrepreneurs are going to have to generate a much larger market.

So far, the best candidates for mass marketing are predictive diagnostic tests that could be conducted on large numbers of healthy people. If an atmosphere can be generated in which none of us feels safe until we have assessed the likelihood that we or our children will develop sundry diseases and disabilities, we will be willing to support this new industry in the style to which it would like to become accustomed. For this re-orientation to happen, researchers will have to stop concentrating on the few, rare conditions that we have hitherto thought of as being "genetic" and begin to implicate conditions for which there is only the slightest evidence of a hereditary component, such as those we have been looking at in chapters 6 and 7.

COMMERCIALIZATION AND CONFLICTS OF INTEREST

As genetic research has become big business, predicted to be one of the major industries of the twenty-first century, conflict-of-interest questions have entered the field. This is not surprising. As "partnerships" have arisen among government, universities, and business interests in other scientific areas, researchers have often found themselves serving several masters. Top nuclear physicists, identified only by their university affiliations, have testified in front of congressional panels about the safety and efficiency of nuclear power, and have later been shown to be paid consultants of the nuclear power industry.[3] In the biomedical area, leading nutritionists on university faculties have had undisclosed affiliations with the food industry[4] and biomedical researchers with drug and pharmaceutical companies.[5] Rarely, however, have these links been as pervasive as they are in the new biotechnology industry.

Since most of the scientists participating in the new genetic research have direct links to biotechnology companies, their research programs and their public scientific pronouncements can affect their economic interests. Every prophecy of a new diagnostic test or medical therapy can affect the standing of biotechnology stocks. Researchers who have investments in these firms or serve as consultants or board members should not simply be accepted as objective scientists.

In 1990 and 1991, six members of genetics panels at the National Academy of Sciences resigned or were asked to sever their ties to private companies owing to concerns about conflict of interest.[6] The Council for Responsible Genetics, a public interest group that keeps track of practical applications of genetics and biotechnology, has devoted an issue of its newsletter, Genewatch, to conflicts of interest in biotechnology. One article mentions the story of Dr. Scheffer Tseng, a researcher at the Harvard-affiliated Massachussets Eye and Ear Infirmary, who was testing a vitamin A ointment to remedy "dry-eye syndrome," a condition that can lead to severe visual disability.[7] While officially on a research fellowship at Harvard, Dr. Tseng was a paid consultant of Spectra Pharmaceutical Inc. of Hopkintown, Massachusetts, and a principal shareholder in this firm, which produced the drug he was testing. In his zeal to prove the value of the drug, he experimented on up to five times the number of people approved in his research protocol, tested it on patients without their consent, and slanted the test results to exaggerate its benefits.[8] Tseng's unjustified claims of success for his treatment were backed up by Kenneth Kenyon, a major Spectra shareholder who was his Harvard supervisor, and by Spectra's founder, Edward Maumenee, a distinguished scientist and emeritus professor at Johns Hopkins University.

Charges were filed with the Massachusetts medical board against Tseng and Kenyon, but were later dismissed on a magistrate's recommendation. According to the Boston Globe, "[The magistrate] found Tseng in violation, but she recommended that the board take into account . . . that 'there was no evidence that Dr. Tseng engaged in fraudulent or unethical behavior' and that he 'is a tireless, dedicated physician.'"[9] This may all be true. The problem is that when scientists have massive financial interests tied up in their research, they will tend to interpret their results as optimistically as possible. Whether Dr. Tseng was deliberately misleading people or simply misleading himself is beside the point for the patients who bought the ointment.

Dr. Tseng's case may be an extreme example, but such dual allegiances permeate the whole research process. In one of the *Genewatch* articles, three social scientists document the extent to which members of biomedical faculties at major U.S. universities are affiliated with for-profit biotechnology companies. They write that "nearly one-third of MIT's biology department consisted of scientists with formal ties to biotechnology companies" and that one-fifth of the faculty of the relevant Harvard departments have similar affiliations. Furthermore, in the National Academy of Sciences—the elite organization of America's top scientists founded by Abraham Lincoln to advise the government on scientific matters—37 percent of the members who are active in the biomedical sciences are known to have commercial ties.[10] For purposes of the *Genewatch* survey, "commercial ties" referred only to academic researchers who were members of scientific advisory boards or standing consultants of biotechnology firms, had managerial responsibilities in a firm, held substantial equity in firms, or served on boards of directors. Considering that only a fraction of the members of biomedical faculties work in fields that overlap the interests of biotechnology firms and that these researchers were able to obtain information on only 549 firms out of a total of 889, it is fair to assume that the level of corporate involvement is higher than the *Genewatch* estimate.

Scientists are routinely called on by the National Institutes of Health (NIH), the National Science Foundation (NSF), and other publicly funded granting agencies to evaluate the quality of research applications and progress reports submitted by their scientific colleagues. They also advise Congress and other public agencies on the safety of new products and technologies. It would be naïve to believe that these scientists' commercial ties do not affect their opinions when it comes to appraising the work either of companies with which they are affiliated or of these companies' actual or potential competitors. In September 1990, the Government Operations Committee of the U.S. Congress issued a report that focused on ten cases of alleged scientific misconduct. One involved an NIH-funded study of t-PA, a new drug designed to prevent blood clots in people who have had heart attacks. The research group, which included several scientists with financial ties to the firm that produces t-PA, evaluated the drug very favorably. Other studies report that t-PA is no more effective than some less expensive drugs, and that it may lead to a higher risk of strokes. As committee member Ted Weiss (D-NY) wrote, "These new results support our concerns that the NIH studies may have been compromised by researchers' bias."[11]

Along with the conflicts of interest they create, such multiple affiliations are bound to limit what and how much scientists communicate about their research results. A glaring example is the case of two Irish researchers who claim they have found a gene mutation associated with hereditary breast cancers. According to an article in *Newsday,* they plan to develop a test to predict these cancers "with nearly 100 percent accuracy." Even this brief article suggests serious problems with the study. However, there is no way to assess its scientific merit, because "the research has not been published, at the request of American Biogenetic Science Inc., an Indiana-based research firm that plans to eventually market the test."[12]

Biotechnology corporations are increasingly funding research in universities. One of the major problems with these ties is that we expect academic research to be open, whereas industrial research is not. The introduction of trade secrets into university laboratories changes the way research is done. Constraints on the open exchange of scientific information also affect the precepts that are being handed on to the next generation of scientists—the graduate students and postdoctoral fellows—about the responsibilities scientists have toward their colleagues, their students, and the public.

In his recent book, Sheldon Krimsky, a professor at Tufts University and the president of the Council for Responsible Genetics, explores the commercial activities of academic scientists and universities in the area of biotechnology.[13] According to Krimsky, not just scientists but many universities themselves hold patents in biotechnology, with the University of California having the largest number and Harvard the second.

In 1980, Harvard University decided not to invest directly in a commercial biotechnology company in which some of its faculty were to be involved, but it has taken out joint patents with faculty members. Boston University has invested upward of eighty-five million dollars in Seragen, a biotechnology firm started by one of its faculty members in Hopkinton, Massachusetts. Boston University is the majority stockholder in Seragen, though its holdings dropped from 92 to around 70 percent when, after seven years, the stock went public in April 1992.[14] Seragen keeps promising innovative drugs, but as of that date it had produced nothing but stock shares and optimism.

In 1981, the Massachusetts General Hospital, one of the chief teaching hospitals affiliated with Harvard Medical School, signed a ten-year, renewable agreement with the German pharmaceutical firm Hoechst AG whereby Hoechst supplies financing for research and in return acquires a considerable degree of control over the research and its uses. Included were provisions that "all manuscripts must be submitted to Hoechst thirty days before submission" and that "Hoechst receives exclusive licenses for all commercially exploitable discoveries."[15] Monsanto Chemical Company has a similar agreement with Washington University in

St. Louis, and such agreements exist between any number of other biotechnology firms and major universities.[16]

Even where universities have not established direct ties with industry, the line can get blurred. In many instances, individual professors have begun to locate private firms on their university campuses. It can be difficult to distinguish projects for which they are receiving federal funds from those that are commercially financed.[17]

These are just a few examples of the current ties between the faculties and governing bodies of research universities and the biotechnology industry. The dangers of such ties have been pointed out in the press, and a series of congressional hearings were held on the subject in the early 1980s. The hearings concerned both the mixing of corporate and federal funds and the use of federal funds for research that was then privately sold. While no one denied the possiblility that such things could, and indeed sometimes did, happen, university administrators hastened to provide assurances that they were quite capable of policing themselves and that the whole problem could be avoided by laying down a few simple guidelines.

Krimsky reports a public dialogue he had in 1979 with David Baltimore, who later became director of the Whitehead Institute, founded by medical supply magnate Edwin C. Whitehead, while working as a professor at MIT and holding major investments in Collaborative Research, a biotechnology firm.[18] According to Krimsky, "Baltimore believed scientists could do it all; serve their universities, serve industry, serve society, and serve themselves. In his view, if done properly this was an all-win situation."[19] Two years after this, while Baltimore was serving on the NIH's Recombinant DNA Advisory Committee, the committee empowered to establish safety guidelines for the conduct of DNA research, the *Boston Globe* reported that he was simultaneously a company director and the single largest individual shareholder of Collaborative Research.[20]

Clearly, for the regulators to be so involved in the industry they are regulating is dangerous. While they may believe themselves capable of keeping all their interests separate, and their right hand independent of what their left is doing, it is hard for the rest of us to be equally confident. Such fragmented allegiances plainly run counter to the public interest.

In the United States, the graduate and postgraduate training of nearly all university-based molecular biologists is paid for by the government, usually the NIH or NSF, through fellowships and training grants. These federal agencies also fund the research most biologists do. In return, the scientists serve on government panels, testify before Congress, or advise the government in other ways. In all these capacities, they are looked upon as disinterested academic researchers. Clearly, this is not accurate in the current situation, in which scientists serve on boards of directors and receive money from commercial firms, personally or to support their research.

Conflicts of interest exist at the highest levels of genetic research. In April 1992, James Watson resigned as director of the National Center for Human Genome Research after the NIH's director, Bernadine Healy, initiated an inquiry into potential conflicts of interest arising from substantial holdings by Watson and members of his family in several biotechnology companies. According to the *New York Times,* "[NIH officials] have said Dr. Watson's decisions as head of the [genome] project could have a substantial effect on the companies in which he has an interest because they are also doing gene sequencing work. Thus they are both competitors with the Government project and potential beneficiaries of its progress."[21]

The situation becomes worse when scientists employed by the government to regulate the development of commercial products or to assess their safety have a financial stake in the success of these products. Such a conflict led to Dr. David T. Kingsbury being forced to resign from his post as assistant director for biological sciences at the National Science Foundation.[22] Dr. Kingsbury chaired the Biotechnology Science Coordinating Committee (BSCC) of the Reagan administration's Office of Science and Technology Policy. At the same time, he was listed on the board of directors of IGB Products, Ltd., a subsidiary of Porton International, a British biotechnology company. Kingsbury also was listed as a consultant with two other California subsidiaries of Porton.

According to a story in the *San Francisco Chronicle,* Kingsbury was providing these companies with expert advice in the development of applications of DNA research for medical diagnoses.[23] Meanwhile, under his direction the BSCC was responsible for developing and coordinating regulatory policy concerning the biotechnology industry for all the federal regulatory agencies, such as the Environmental Protection Agency and the Food and Drug Administration.[24]

Many people have come to question whether the NIH should be the government agency charged with regulating research in the biomedical sciences when its principal function is to promote such research. Yet, since the late 1970s, the Recombinant DNA Advisory Committee (RAC) of the NIH has had the responsibility of overseeing the safety of government-funded DNA research. (Though industrial researchers have agreed to conform to the RAC's guidelines, their compliance is not monitored by any federal agency.)

Conflicts of interest are not new in science. The U.S. Congress is always grappling with questions of how to regulate the transfer of information and personnel between publicly funded research institutions and privately owned commercial firms without unduly infringing on scientific freedom or entrepreneurial gusto. Yet the relationship between biomedical researchers and for-profit ventures has never before been so extensive and therefore so threatening to the public interest and damaging to the public trust.

In his article in *Genewatch,* Representative Ted Weiss argued that perhaps the government should institute regulations that forbid researchers supported by the National Institutes of Health (NIH) "from owning stock or equity in a company that makes the drug they are evaluating with federal funds." He further suggested that "at a minimum, NIH-funded scientists should be required to disclose their financial ties to relevant companies every time they present their research results orally or in writing."[25] These suggestions seem modest indeed.

OWNING THE GENOME

We need to ask why knowledge that has been developed at public expense is routinely turned over to commercial firms at the point at which it can be made to yield biomedical benefits. Why should the public not have access, at cost, to the results flowing from knowledge its taxes have paid to generate?

At present, the opposite is happening. A drama is unfolding in which not only scientists whose research is supported at public expense, but the NIH itself, are entering the commercial arena. In October 1991, *Science* magazine published a story about a scientist named Craig Venter who, with colleagues at the National Institutes of Health, is attempting to identify all the functional DNA sequences in the human brain.[26] These scientists are taking sequences of messenger RNA found in human brain cells and transcribing them back into DNA.

By determining the base sequences of such chunks of "complementary DNA" (complementary, that is, to the messenger RNA), Venter and his colleagues hope to form a library of what they call "expressed sequence tags." These are base sequences that presumably get translated into the proteins that function in the brain. The scientists hope that these "tags" will enable them to detect all the DNA sequences that are used in the brain. To put it differently, they are trying to identify a bunch of genes, without knowing their functions or where on the chromosomes they occur.[27]

Here the plot thickens. In the summer of 1991, at a congressional briefing on the Human Genome Project, Venter mentioned that the NIH was filing patent applications on all these "tags," which he is identifying at the rate of 50 to 150 per day. The idea is that by patenting the sequences now the NIH will "own" them, if and when that knowledge can be exploited commercially.

According to the *Science* story, "James Watson, who directs the genome project at NIH . . . explode[d] and denounce[d] the plan as 'sheer lunacy.' With the advent of automated sequencing machines, 'virtually any monkey' can do what Venter's group is doing, said Watson."[28] This might sound as if Watson objects to patenting DNA sequences, but in fact he is only objecting to Venter patenting them at such an early stage, before anything is known about their functions. Watson feels that it is wrong to patent random sequences, not that it would be wrong to patent genes.

No one here is fighting for the purity of research, or insisting that scientists avoid commercial entanglements. They are just arguing about how best to make the fruits of their research commercially viable. Bernadine Healy, the director of the NIH, says that, although the sequence tags are now meaningless pieces of DNA, the NIH had to seek patents in order to be able to publish the sequences without them falling into the public domain. She argues that the lack of such protection might discourage biotechnology companies from identifying the function of these sequences and using the information to develop products.[29] Opponents, even within the biotechnology industry, say that, on the contrary, patenting at this stage is likely to limit both scientific and enterpreneurial interest. As David Botstein, a genome project researcher, says: "No one profits from this, not science, not the biotech industry, not American competitiveness."[30]

In 1992, after the NIH had applied for patents for nearly 2,000 more gene tags, Venter, with his research team, left to head a newly formed $70-million research center funded by a venture capital group called the Healthcare Investment Corporation.[31] Meanwhile, Healy has said that NIH would be willing to forego patents on gene tags if there could be international agreement that this "will not impede the ability to obtain adequate patent protection" for future products.[32] However, when the U.S. Patent and Trademark Office rejected NIH's patent application in August 1992, Healy let it be known that she would appeal the decision and expected to win.[33]

British scientists are doing research similar to Venter's, but the Medical Research Council (MRC), Britain's equivalent of the NIH, has announced that it will not patent random DNA fragments. Instead, the

MRC is keeping the composition of the fragments secret and intends to sell access to this information to biotechnology companies that want to use it to develop products.[34]

In all this, it is important not to lose sight of the fact that the pieces of DNA being sequenced are part of our bodies; they are not being invented by these researchers. If the base sequences of this DNA can be patented, rather than remaining in the public domain, the rights to the commercial use of these sequences will belong to the NIH or to companies that buy them from the MRC. In the end, consumers will be the losers. They will first pay the costs of the research and patenting with their taxes, then pay prices inflated by monopolies.

WHAT TO DO?

Clearly research on DNA is not just about understanding nature or helping suffering humanity. The scientific exploration of our genes, its potential medical benefits, and corporate interests have become inextricable.

It may be a new idea for us to regard a scientist's or physician's recommendations as skeptically as we do those of other kinds of salespeople, but we need to do just that. To keep control of this tangled situation, we must learn what questions to ask, and we must not accept that the answers are too complicated for us to understand. We need to be sufficiently well informed to be able to evaluate critically what the "experts" tell us, so that we can make our own judgments about what tests and what information are likely to benefit us.

For most of us this represents an entirely new way to conceptualize our relationship to scientists, physicians, and the health care system. We will not be able to reshape our attitudes by ourselves, but fortunately a few consumer and public interest organizations already concern themselves with the individual and societal implications of the new genetic technologies, and others are likely to be formed as more people come to appreciate the changes that are taking place.

It took several decades for the public to become aware of the dangers posed by the petrochemical and atomic industries, and for citizens' groups to mobilize and work to limit and, when possible, to eliminate their risks. By and large, responses to genetic manipulation and to developments in biotechnology have come more promptly.

In the late 1970s, citizens' groups in towns and urban areas where genetic engineering was being done drafted ordinances to regulate practices in biotechnology firms and university-based research laboratories.

These groups tried to minimize the risk that biohazards might escape into the surrounding areas. At present, a number of public interest and other nonprofit organizations provide information and lobby legislators about relevant issues. (On pages 197–198, you will find a list of some organizations dedicated to providing information and organizing skills.)

Obviously, we need stricter regulation of conflicts of interest. Medical geneticists and genetic counselors should have to disclose their corporate connections and their financial interests in companies that are developing or administering genetic tests. Clients should know when the person who is informing them about the pros and cons of a particular test or drug has a financial stake in their taking it. Perhaps we should ask that physicians exhibit certificates of their consultancies, equities, and memberships on boards of directors on the walls of their offices, along with the usual diplomas and certificates of membership in professional associations.

Similarly, just as scientists now list their academic affiliations on their publications or when they make public pronouncements, they should be required to disclose their corporate connections. Scientists applying for grants to support their research are usually asked to list all other sources of research support. But that requirement does not address the issue I have in mind, because at present they need not divulge their investments and other financial interests, as long as these do not specifically support their research. And yet, scientific colleagues, the public, and the government bodies that receive scientists' advice have a right to know about the interests that may bias that advice.

We must not let decisions in matters that intimately affect us be determined by experts whose impartiality is far from clear. Of course, we must be able to choose whether or not to use new technologies. But more important, we must become part of the processes that determine what technologies should be developed. This is true in the medical and health fields, and equally urgent when we come to the subjects of genetic discrimination and forensic genetics, which we will consider next.

. . . .

TEN

GENETIC DISCRIMINATION: EDUCATION, EMPLOYMENT, AND INSURANCE

GENETIC TESTING AND THE SCHOOLS

More children in the United States leave school functionally illiterate and unable to do simple arithmetic than in any other technologically developed country, and in many of the less developed ones as well. As schools are asked to take on an ever-increasing role in socializing children to function in society, they find themselves under constant criticism for failing to educate these children to even minimal standards.

Over the past few decades, the educational establishment has increasingly countered this mounting dissatisfaction by looking for problems in the children, rather than facing the problems in the learning environment or the broader society. Schools have developed long lists of diagnostic labels for so-called learning problems, which get interpreted as though the label itself provided information about the reasons a child is not doing well in school. Educators feel relieved if they can somehow attribute a child's problems to "underlying" biological causes, even when they cannot point to specific biological evidence. In this climate, any test, however preliminary, gets used.

In their book *Dangerous Diagnostics,* the social scientists Dorothy Nelkin and Laurence Tancredi discuss some of the ways schools and teachers are trying to meet their problems by expanding the list of available diagnostic tests and labels.[1] Many of the newer tests try to link supposed brain dysfunctions with "learning disabilities," variously labeled "dyslexia," "attention deficit disorder," and "defective short- or long-term memory." These tests are increasingly interpreted and used as though they could reveal—and even predict—causal, genetic relationships, though in fact they can only correlate physiological phenomena with the various ways in which children perform. Of course, children do experience learning problems, and such problems may sometimes be related to biological dysfunctions. But I distrust the obvious relief with which some teachers, school administrators, and parents locate the source of such problems within the children's genes and brains.

A psychologist, Frank Vellutino, reports in *Scientific American* that a gene has been localized "on chromosome 15 in members of families in which there is a history of reading disability." He adds that "once a gene that may be responsible for a specific attribute has been localized on a specific chromosome, geneticists are in a position to find the mechanisms whereby the gene gives rise to the attribute." Therefore this finding "could be a significant breakthrough in the study of dyslexia, but it has not yet been replicated."[2] As we have seen, there is an enormous distance from a gene and the protein in whose synthesis it is implicated to a complex behavior like a "learning disability." Psychologists and educators need to understand this and to stop expecting practical benefits from oversimplified genetic correlations.

Such correlations can be used to take the educational system off the hook, but cannot help students to learn more effectively or teachers to teach Johnny to read. As we have seen before, experts often act as though rooting a problem in an identifiable genetic "defect" solves the problem itself, but this is simply not true. However, as genetic and other biologically based tests become part of the standard testing apparatus in schools, teachers and parents will come to rely increasingly on seemingly objective labels that often have little diagnostic and even less predictive value. Nelkin and Tancredi point out that "over time, routine use obscures the uncertainties inherent in [genetic] tests [and leaves] their underlying assumptions unquestioned."[3]

The current love affair with predictive tests for "learning disabilities" sets up the potential for discrimination in the future as well as the present. Norms create deviance, and an "abnormal result" on a biological or genetic test, though it does not blame the child, stigmatizes her or him and projects that stigma into the future. I.Q. tests, which were designed to pinpoint those areas in which a child needed special attention, have come to be used as measures of "intelligence," marking people with a single number that is supposed to represent not only their current but their "potential" abilities.

Diagnostic labels can affect a child's self-image and his or her relationships in school and at home. They also become part of that child's "file," the growing body of data that follow her or him from school to school and job to job. Nelkin and Tancredi put it this way:

129

The use of these diagnostic techniques has substantial social force beyond the educational context. The school system has contact with most children in the society, and is traditionally responsible for assessing, categorizing and channeling them toward future roles. . . . School professionals . . . transmit their evaluations to other institutions to help identify who is genetically constituted to assume certain types of jobs. Thus, diagnostic technologies not only help schools meet their own internal needs, they also empower schools in their role as gatekeepers for the larger society.[4]

At present, in most circumstances, parents have the right to refuse to let schools administer special tests to their children. (Mandatory drug testing is an obvious exception.) But this option can create a double bind like the one we encountered when looking at prenatal tests. Once tests become the accepted norm, persons who refuse them cast doubt on their probity, if not their mental competence. Also, children often need to have a "diagnosis" in order to have access to special resources, such as tutoring, smaller classes, resource rooms, and the like. So, parents are forced to have their children tested to get practical help.

Nelkin and Tancredi suggest that genetic testing in the schools could become mandatory if enough people come to believe that a specific genetic condition affects behavior or the ability to learn. This is especially likely if people can be persuaded that such new information will help relieve behavior problems and so benefit both the affected children and their classmates.

It is all too easy for genes to take on a life of their own. Genetic "learning disabilities" are a stigma not only for the child who has them but for all the relatives and descendants of that child. They can be used to show why poor children do not do well in school and to explain why their families got to be poor in the first place and will continue to be poor in the future. As usual, there will be plenty of rhetoric about how educators are only trying to identify children who may need special attention, but in a world of shrinking school budgets this is pure fantasy. The tests will serve as an explanation and an excuse, getting schools and society off the hook by placing the blame on the children's unchangeable genes.

GENETIC DISCRIMINATION IN THE WORKPLACE

The dangers of genetic testing do not end when a person leaves school. In workplaces genetic tests are used to screen workers and to monitor them. *Genetic screening* is done in order to find out whether job applicants are more likely than an "average person" to develop medical conditions that are thought to be inherited biologically and that might reduce their effectiveness as workers. Such screening is likely to be done just once. *Genetic monitoring,* by contrast, is done periodically to find out whether some chemical or other hazard in the workplace is altering the chromosomes or genes (DNA) of the workers.

Employers usually shy away from genetic monitoring, since it tends to implicate workplace chemicals as the source of harm. However, employers have economic reasons to screen prospective or actual workers in order to keep people with potential health problems out of the workplace. Since businesses are run to make a profit, employers try to minimize labor costs. Such costs include time spent training employees and paying employees' benefits, such as health and disability insurance. Nearly all but the smallest firms offer health insurance as part of their benefits program, and large businesses have begun to cover the costs of their employees' health care themselves rather than contracting with insurance companies. According to Nelkin and Tancredi, "employment benefits exceeded 39 percent of total payroll costs in 1986, and over 21.1 percent of that total went for medical benefits. . . . Where companies buy insurance from commercial carriers, premiums are experience rated [and] higher claims mean higher costs."[5] So, insurance costs are one reason that employers may try to minimize health insurance claims by employees.

It is also in the interest of employers not to spend time training employees whom they will lose through sickness or death, as well as to reduce the cost of disability benefits. However, while it is costly to have workers become ill, it is also costly to keep workplaces uncontaminated by toxic chemicals used in the manufacturing process, and to take the various safety precautions that may be necessary to preserve workers' health and well-being. Employers will therefore want to use tests that promise to predict the future health of prospective employees, in order to weed out job applicants who might be unusually sensitive to hazards in the workplace.

Employers have used the concept of the "accident-prone" worker to shift responsibility for industrial accidents onto the people who are injured.[6] For example, though there are consistently more accidents on the graveyard shift, such accidents are often blamed on the carelessness of individual workers rather than on the difficulty of working through the night. By the same token, many employers now embrace the concept of genetic "hypersusceptibility" to explain why some workers respond to lower levels of dusts or other contaminants than the "average worker" does.

Autoworker mounting roofracks, assembly line, Detroit, Michigan. Most industrial accidents and injuries occur when the work process is not designed to minimize strain on the joints and muscles of the workers. (Photograph © Earl Dotter.)

In a 1984 article I wrote with Mary Sue Henifin, an attorney trained in public health, we discussed problems inherent in the notion of "hypersusceptibility" to contaminants.[7] For one thing, it is difficult to decide on appropriate criteria: How low (or high) must the level of exposure be before someone who is harmed is labeled "hypersusceptible"?

Another point is that the same industrial chemical or other toxic agent can provoke acute reactions in some people in the short term, while other workers may slowly develop chronic conditions, such as cancers, without ever exhibiting an immediate, acute reaction. Most screening tests in current use detect only those workers likely to develop the short-term reactions, but not those who may experience long-term effects. However, molecular biologists are now laying the groundwork for genetic tests said to predict "tendencies" for the more slowly developing, chronic conditions. When such tests become available, they will be even less reliable in their predictions than the tests that currently promise to identify "hypersusceptible" individuals, since so many factors can contribute to chronic conditions.

Predictive tests should never be used to screen out workers who are considered "hypersusceptible," whether to short- or long-term effects. Whatever their individual "susceptibility" may be, all workers who are exposed to toxic agents risk developing a chronic condition sooner or later.

For example, formaldehyde is a common contaminant in the plastics industry, among others. It can evoke acute allergic reactions such as asthma or skin rashes in some people, and is also known to be a human carcinogen. Once predictive tests are available, employers will want to screen out workers who might have allergic reactions or be at greater than average risk of developing cancer. This could then allow them to expose everyone else to high levels of this carcinogen. It would be far better to find a chemical substitute for formaldehyde or change the procedures so as to lessen the risks of both types of effects for all the workers, but to a cost-conscious industry that sounds like a more expensive proposition.

Genetic tests, far from being safety measures, can lead to a relaxing of existing precautions. However, the scientists involved in developing such tests often stress only the potential benefits. An opinion piece in the British scientific weekly *Nature,* which criticizes those of us who warn that the Human Genome Initiative will increase the potential for discrimination, asks editorially: "Would it not be profitable to keep the plants [which use vinyl chloride, a carcinogen that produces liver cancers and to which some people are supposedly 'hypersusceptible'] going and to use part of the economic wealth created to compensate those with the bad luck to be susceptible?"[8] Such questions can only be asked in an academic ivory tower. Of course such a policy would make economic sense, just as it would make economic sense to share the profits of mechanization with the workers who are replaced by machines, but that is not how our society works.

It is encouraging that in October 1991, the Council on Ethical and Judicial Affairs of the American Medical Association published a set of guidelines "to help physicians assess when their participation in genetic testing [by employers] is appropriate and does not result in unwarranted discrimination against individuals with disabilities."[9] In its statement, the Council stresses that genetic tests are poor at predicting diseases and even poorer at predicting whether a specific health problem will interfere with an individual's work performance. The Council therefore considers it inappropriate for physicians to participate in testing that assesses anything beyond a worker's ability to perform the actual tasks that are part of the job. The Council also states categorically that "testing must not be performed without the informed consent of the employee or applicant for employment."

Although a number of genetic traits have been implicated as possible predictors of "hypersusceptibility" and others will be suggested as more genetic tests become available, there are no adequate data to link any genetic trait to a specific industrial disease. What is more, little, if any, research is being done to explore the parameters within which such predictions might conceivably be valid.[10] Despite this, the Office of Technology Assessment (OTA) of the U.S. Congress published a report in 1990 that includes a table entitled, "Identification and Quantification of Genetic Factors Affecting Susceptibility to Environmental Agents."[11] This table lists twenty-seven "high-risk groups" of people who are described as genetically "hypersusceptible" to environmental contaminants. The authors tone down that claim by stating that these groups only "may be" at risk, but why publish such a table at all when it is based on little, if any, reliable information?

According to surveys the OTA conducted in 1982 and 1989, few employers are using predictive genetic tests. But these surveys had only a 6 percent response rate. Even if the survey results are representative, the process of testing is still new and relatively expensive, in some cases running into thousands of dollars.[12] Many an employer who does not use tests now might act rather differently if prices come down.

Larry Gostin, an attorney and the executive director of the American Society for Law and Medicine, predicts that "market forces may be the single greatest factor motivating genetic testing." He points out that "market researchers project that U.S. genetic test sales will reach several hundred million dollars before the decade's end"[13] and that this will lower the cost of testing and encourage its use by employers and insurance companies. As some insurers or employers become more sophisticated in their use of genetic predictions, this will put economic pressure on others to do the same. When many more tests are available and they are cheaper to administer, employers are likely to try to use them to screen out potentially costly workers, whether or not the tests are reliable.

As always, the discriminatory potential will not be felt equally in all jobs or by all applicants. A highly skilled person with unique qualifications will be less likely to be screened out than an applicant for a more routine job, for which many others could be hired. So here as elsewhere, the least powerful segments of society are most likely to be exposed to discrimination.

Although preemployment genetic discrimination is not yet an everyday occurrence, Paul Billings and his colleagues have already come across it in their preliminary research. One of their respondents listed *Charcot-Marie-Tooth Disease* (*CMT*), which is a heritable neuromuscular disorder known for its highly variable clinical manifestations, on a preemployment form. The interviewer asked her what CMT was, looked it up in a medical book, and did not hire her.[14] In another instance, a healthy young man who carries one allele associated with a recessive condition called *Gaucher disease* was not permitted to enlist in the Air Force, though this condition is never manifested in the carrier state. Billings's relatively informal survey has turned up two instances of genetic discrimination in employment, even though few employers are yet geared up to administer genetic tests. We may expect to see many more cases when genetic testing becomes widespread, unless legal safeguards are put into place.

If genetic tests could predict a person's health status with fair accuracy, and if employers did not use them primarily to save expenses of training people who might get sick, predictive tests might benefit some workers. But these are very big "ifs." Since the test results can jeopardize a worker's future employment possibilities, genetic tests are not likely to meet the health needs of workers or further their welfare. This is why during the past decade, trade unions and their allies have been outspoken in their opposition to the use of genetic screening tests. They have noted that tests that emphasize inborn genetic differences as the causes of potential disabilities are by their very nature discriminatory, because they sort people on the basis of factors that are beyond their control. At present there are no safeguards to limit the future use of such information. Also, in weighing the potential benefits of such testing it is often assumed that workers who are denied employment in one kind of job or industry because of a given physical condition can turn elsewhere for work, but this is far from true in the present state of the U.S. economy.

Unless there are sufficient numbers of adequate jobs for everyone who needs or wants to work, so that people can find not only a job but one that is suitable, and until workers and management jointly have exhausted the possibilities to reduce workplace hazards, genetic screening will threaten the health of workers, not improve it. At present, predictive tests are more likely to divert attention from the unnecessarily dismal conditions of most workplaces and so only benefit management.[15] Even when workplace conditions are improved, the results of such tests must be privileged information that is available only to the tested worker.

We will be looking at such privacy issues in the next chapter, but first let us see to what extent such legislative measures as the Americans with Disabilities Act of 1990 (ADA) can reduce the risk of genetic discrimination in the workplace.

MEASURES TO COUNTER EMPLOYMENT-RELATED
DISCRIMINATION

In its report, *Genetic Monitoring and Screening in the Workplace,* the Office of Technology Assessment (OTA) of the U.S. Congress states that without protective contracts or legislation an employer has "virtually unlimited authority to terminate the employment relationship at any time. . . . [This] includes the right to refuse to hire an individual because of a *perceived* physical inability to perform the job and the right to terminate employment because of a *belief* that the employee is no longer able to perform adequately" (italics mine). The report goes on to point out that "even if test results were inaccurate or unreliable, the employer would be protected in basing employment decisions on them."[16]

Though courts have begun to create precedents based on antidiscrimination law that limit this "right" of employers, genetic testing is so new that as yet there is no relevant body of judicial attitudes or opinions to regulate its use. Meanwhile, the OTA report estimates that the tests that are at present available to detect simple Mendelian traits or inherited chromosomal aberrations could affect some 800,000 people in this country, while "potential future tests" could affect some ninety million people.[17] Included are tests for so-called genetic conditions that confer a "tendency" or "predisposition" to develop hypertension, dyslexia, cancer, and seven other common physical or behavioral disabilities.

Various federal laws may provide some protection for workers by restricting the right of employers to impose mandatory genetic tests, to use test results to discriminate against workers, or to breach confidentiality. These include the Occupational Safety and Health (OSH) Act which set up the Occupational Safety and Health Administration (OSHA), Title VII of the Civil Rights Act of 1964, the Rehabilitation Act of 1973, the National Labor Relations Act, and the Americans with Disabilities Act (ADA) of 1990. Unfortunately, there are problems with all of these.

To come under the protection of the Rehabilitation Act, which prohibits employers who receive federal contracts or other federal subsidies from discriminating against persons with disabilities, employees must be able to prove that their genetic condition constitutes a genuine impairment, but that they are otherwise qualified to do the job. If they can do that, the employer must provide them with "reasonable accommodations." The ADA extends this protection to the private sector and, by 1994, is supposed to cover all employers with fifteen or more employees. The ADA stipulates that preemployment medical examinations or inquiries may be used only to determine an applicant's ability to perform the actual job in question. In other words, an applicant cannot be disqualified unless the condition interferes with specific tasks required as part of his or her job. This is very important for people with all kinds of disabilities. However, since the ADA does not make specific reference to inherited conditions or to tests designed to detect or predict them, it is not clear how the courts will apply these provisions to now-healthy people said to have a "tendency" to develop a disability at some undefined future time.

The OTA report suggests that OSHA is the most likely federal agency to monitor genetic testing in the workplace, since it has dealt with other biological tests in the past, but it is not clear what form its monitoring will take. In fact, the OSH Act takes no position on genetic testing and is not concerned with protecting employment rights. Besides, OSHA has been so chronically understaffed that it is unrealistic to expect it to monitor yet another set of workplace practices effectively.

Individual states are likely to address the issue of genetic discrimination by virtue of their involvement with workers' compensation programs. The OTA found that in 1983, four states (Florida, Louisiana, North Carolina, and New Jersey) had statutes restricting the use of genetic information in employment decisions, but all except New Jersey mentioned only genetic screening for the sickle-cell gene. Only New Jersey has a law banning employment discrimination on the basis of genetic tests.

Gostin has done a detailed analysis of the extent to which the provisions of the ADA are likely to be useful to prevent genetic discrimination. He believes that "persons currently disabled by a genetic disease are undoubtedly covered under the ADA," but since the courts define "disability" as a "'substantial' limitation of one or more life activities," it is not clear whether a genetic condition that can be argued not to cause "substantial" impairment will count as a disability.[18]

As far as predictive diagnoses are concerned, it is not clear whether a currently healthy person who has been shown to carry the allele implicated in Huntington disease, and who will therefore become disabled at some unspecified future time, can be classified as "disabled" within the meaning of the ADA and is therefore protected against genetic discrimination. Gostin thinks the ADA covers the "healthy ill" and that such people would be protected in the same way as people who test positive for HIV are protected even before they develop symptoms of AIDS. The same reasoning should also apply to people who have genetic tests that show them to be "predisposed" or "susceptible" to develop heart disease or cancer.

The ADA was intended to protect anyone who has or is *perceived* to have a disability. Therefore, Gostin argues, it would betray the spirit of the Act to interpret it as permitting discrimination against individuals who have had genetic tests that predict they or their children may become disabled, simply because they have not yet experienced the predicted disability. In enacting the ADA, Congress ruled specifically that "an inquiry or medical examination that is not job-related serves no legitimate employer purpose, but simply serves to stigmatize the person with a disability."[19] Congress also made it clear that, even when there is reason to think that applicants may become too ill to work in the future, employers cannot cite training costs as valid reasons to discriminate against them in hiring or other employment decisions.

Employers also cannot cite increased costs of health care or insurance benefits. Gostin believes that these safeguards will be effective in limiting the ability of employers to use genetic tests before offering someone a job. The ADA allows employers to require medical examinations once a person has been hired, but only if the same tests are given to every new employee and if all medical information is kept confidential. Even then, the examination must be relevant to the job and justifiable as serving business interests.

These measures appear to offer protection against genetic discrimination. However, employers are exempted from some of these provisions if they themselves also serve as insurers, which increasing numbers of large employers do. As we will see shortly, the antidiscrimination provisions do not apply to insurers.

Despite his generally optimistic reading of the ADA, Gostin points out that while the Act prohibits "discrimination based upon past disability ('record of impairment'), current disability ('impairment'), or perception of disability ('regarded' as impaired) . . . [it] is silent about discrimination based upon future disability."[20] Gostin suggests that the ADA could easily be strengthened if its definition of disability were broadened to include: "having a genetic or other medically identified potential of, or predisposition toward, an impairment."[21] This change would be extremely important. At present, if a healthy person is predicted to develop a condition such as Huntington disease, a court could rule that she or he is not currently impaired and therefore not protected by the ADA.

There are also other problems.. An employer does not have to give reasons why a particular applicant is not hired. If employers can conduct predictive or other medical tests as part of the preemployment process, they may use medical information from these tests to make hiring decisions, but claim to have based their decisions on reasons unrelated to health. It is easy to imagine how this practice could be extended to genetic conditions, so it is important that the ADA prohibit employers from requiring medical examinations or inquiries before hiring a job applicant, except as they relate to that person's present ability to do a specific job. (Drug testing is excepted because drug tests are not considered medical examinations.) Yet, here again, these restrictions will not apply to employers who self-insure. To be effective in preventing genetic discrimination, state laws need to be directed more specifically against discrimination based on predictions of genetically mediated disabilities that might become manifest at some unspecifiable time in the future.

There is no question that the U.S. Congress and state legislatures could improve this situation by passing laws that specifically prohibit genetic discrimination in hiring and employment, but we must recognize that laws can address only part of the problem. Although civil rights legislation in other areas has been important and has curbed many abuses, civil rights laws have not ended discrimination. When people are oppressed, and have few resources at their disposal, they are often unable to resort to legal remedies. Furthermore, prospective employers are not accountable to people they interview for job openings. At a time when there is no shortage of applicants, employers will have no difficulty concealing discriminatory reasons for their decisions about hiring and firing.

As for the right to refuse tests, such a right has only a limited value. Job applicants or employees are at an even greater disadvantage in dealing with employers than prospective parents are with physicians or parents are with school authorities. Even if job applicants or workers have the right to refuse predictive tests, they may not know they have that right. If they do know, they may not be in a position to exercise it. Even though it seems that most medical geneticists favor voluntary over mandatory testing in the workplace,[22] the distinction may be largely academic. As we will see in the next chapter, someone applying for a job can easily be made to sign away rights to privacy and confidentiality.

Because of the impact genetic discrimination could have on people's health and well-being, people need access to education and information about these issues. Unions and public interest groups must insist that protective legislation and contract language are enforced. Model laws must be developed to counter this new form of discrimination. Mechanisms must be put into place that regulate the ways in which decisions are made about what scientific research is appropriate in this area and the rate at which it should be done. Also, as scientific capabilities to offer individuals fateful predictions of uncertain validity keep expanding, we must decide how best to anticipate and minimize the economic and social damage such predictions will produce.

GENETIC DISCRIMINATION IN INSURANCE

In his preface to *The Doctor's Dilemma*, George Bernard Shaw satirized the situation of the medical profession, which is supposedly in the business of healing, but derives its livelihood from sickness. To quote:

> It is not the fault of our doctors that the medical service of the community . . . is a murderous absurdity. That any sane nation, having observed that you could provide for the supply of bread by giving bakers a pecuniary interest in baking for you, should go on to give a surgeon a pecuniary interest in cutting off your leg, is enough to make one despair of political humanity. But that is precisely what we have done. And the more appalling the mutilation, the more the mutilator is paid. He who corrects the ingrowing toenail receives a few shillings: he who cuts your inside out receives hundreds of guineas, except when he does it to a poor person for practice. . . . I cannot knock my shins severely without forcing on some surgeon the difficult question, "Could I not make a better use of a pocketful of guineas than this man is making of his leg? Could he not write as well—or even better—on one leg than on two? And the guineas would make all the difference in the world to me just now. My wife—my pretty ones—the leg may mortify—it is always safer to operate—he will be well in a fortnight—artificial legs are now so well made that they are really better than natural ones— evolution is towards motors and leglessness, &c., &c., &c.[23]

The for-profit health insurance industry raises this contradiction by several notches. Insurance companies make money only so long as people pay more to buy health insurance than it costs the insurance company when these people feel so ill that they consult a physician. So, to make a healthy profit, insurance companies should sell most of their insurance to people who won't get sick.

In the real world, insurers get around this dilemma by using actuarial and other sources of information to estimate the probability that someone will get sick on the basis of his or her membership in a specific group, indentified by age, sex, profession, and other criteria. The insurance industry therefore is in the business of discrimination, since it sorts people into groups on the basis of criteria over which they have no control, and then sets their premiums on the basis of that group's statistical risk of developing specific illnesses. The idea is to make money despite the fact that insured people will get ill and the insurer may have to pay out quite a lot before some of them get well or die.

Insurance companies also use various means to identify those applicants who are more likely to develop a medical condition than their group membership might suggest. The use of this information to determine a person's insurability is known as "underwriting." If the insurance company has reason to believe someone will turn out to be a costly client, it may charge that person higher premiums. If the company believes a person is likely to develop a specific condition, it can also refuse to insure him or her for that condition, or cancel the relevant coverage. Already in 1989, pediatrician and epidemiologist Neil Holtzman was able to list nine conditions, including sickle-cell anemia, arteriosclerosis, Huntington disease, type 1 diabetes, and Down syndrome, for which insurers had denied medical or disability insurance, and six others for which they granted only conditional or partial coverage.[24]

Insurers want to have as much and as accurate predictive information as they can lawfully get before they insure anyone. By the same token, it is in a client's interest to withhold any information that makes her or him appear other than "average."

Along with individual insurance policies such as I have been discussing, insurance companies also issue group policies to businesses. The price of these policies is set on the basis of the health care costs the business has experienced in previous years. This process is called "experience-rating." Employees in such businesses are eligible for the group plan without being tested individually, unless they admit they have certain, specific medical conditions. If they have an excludable health problem, they must buy an individual insurance policy, which is usually written so that it excludes coverage for their "preexisting" condition.

Insurance companies point out that, owing to the way insurance currently works, people who pay their premiums without getting sick pay for the medical expenses of the ill. Therefore, the companies say, it is only fair that people who are already sick, or who are more likely to get sick than the "average" person, should pay higher premiums, rather than making other people pay their health care costs. They also argue that it should be within their rights to renegotiate contracts so as to eliminate certain coverages if an insured person's health status changes, as for example if the person tests positive for HIV. Such rewriting defeats the purpose of health insurance. If the terms of a policy change as soon as one gets sick or is predicted to become sick, the insurance is worthless.

If exclusionary practices are allowed, the existence of supposedly predictive tests for an increasing number of common conditions such as cancer, high blood pressure, or diabetes will surely exclude people outright or force them to pay more for insurance. Not only will these tests permit a glimpse into someone's distant future (however fogged that glimpse may be) but they may suggest something about the health of that person's future children, who might be covered by the same insurance policy. Widespread use of predictive genetic tests is bound to exacerbate the injustices inherent in for-profit health insurance. More people will join the pool of "uninsurables" who must rely on public social insurance, which we pay for out of social security and other taxes. Clearly, nothing short of universal coverage by a national health plan similar to those in Canada and the European nations will remedy this situation. Such a plan must guarantee access to health care for everyone, irrespective of their present or future health status and their ability to pay for health insurance or medical care.

Insurers are not yet making extensive use of predictive genetic tests but, like employers, they are likely to use them once such tests become less expensive than they are at present. Legislation and organizing are needed to forbid this form of discrimination. Let us remember that it took years of court battles based on prohibitions against sex discrimination in employment to stop employers and insurers from using actuarial grounds to pay women lower retirement benefits than men, which they had done on the grounds that, statistically, women live longer. The same effort will be required here.

Already there are documented instances of genetic discrimination by insurance companies. Just as Paul Billings and his colleagues uncovered preemployment discrimination, they also came across a man who was denied automobile insurance because he had a predictive genetic diagnosis of Charcot-Marie-Tooth disease (CMT), although he had never had an automobile accident in twenty years of driving and his physician certified that he had no symptoms of this condition.[25] Another person with CMT could not buy life insurance, despite the fact that CMT does not affect life expectancy.

In two instances, "women carrying fetuses, which had been diagnosed as having genetic disorders, decided to continue their pregnancies. They then had to fight to retain full insurance coverage for the future care of their babies."[26] Billings and his colleagues also cite insurance discrimination against someone who was still healthy, but was known to carry the allele involved in Huntington disease,[27] and Gostin cites the instance of someone who was not allowed to buy insurance because he had a diagnosis of *hemochromatosis,* which is a controllable malfunction of iron metabolism.[28] In a "60 Minutes" interview in May 1992, Jamie Stephenson described having her entire family's health insurance cancelled after two of her children were diagnosed to have *fragile X syndrome,* a variable condition involving mental retardation.[29]

Insurance practices such as exempting "preexisting conditions," limiting coverage, charging higher premiums for higher perceived risks, or changing existing insurance policies are bound to have a serious impact on people said to have a genetic "predisposition" to develop cancer or some other condition. As Gostin points out: "If insurers have actuarial data demonstrating a likelihood of future illness, they can limit coverage [of that illness]. More worrisome would be a decision by an insurer to view a genetic predisposition as a preexisting condition." He adds that the greater the predictive value of tests gets to be, "the more likely . . . that insurers will regard the condition as uninsurable or preexisting."[30] This would not be unlike the reaction some insurance companies have had to HIV infection, which has been to require HIV testing and to consider persons who test positive uninsurable.

The only limitation the Americans with Disabilities Act places on insurance companies is that it does not allow them to refuse coverage to someone for other health conditions because that person has a specific genetic prediction. They can refuse, cap, or limit insurance, but only for the predicted condition. Obviously, this can have a devastating effect on someone who becomes ill and is denied insurance. Already, insurers have rewritten policies to exclude coverage for AIDS after a policyholder has become ill with AIDS-related conditions, thus denying him or her health insurance at the time when it is most needed. Some states have passed legislation prohibiting this practice, but most allow it.

The ADA forbids employers to ask non-work-related questions about their employees' health. However, now that large companies increasingly serve as their own insurance carriers, they can ask for information as insurers that the ADA would prevent them from asking in their role of employer. Also, as insurers, they have access to the Medical Information Bureau in Westwood, Massachusetts, a centralized data bank of health information for all of North America. While they are not supposed to use this information to make employment decisions, once they have it there is no way to control the ways in which they may use it.

In September 1991, the California legislature passed an amendment to the California Civil Rights Act that provided for an eight-year ban on the use of genetic information by health insurers and employers and on the use of such information to limit access to group life and disability insurance. Although this amendment had bipartisan support, Governor Pete Wilson vetoed it.

Also in 1991, Wisconsin passed legislation forbidding any person or organization to require an individual to take a genetic test or reveal whether she or he has taken such a test. Moreover, the results of such a genetic test for any individual or family member cannot be a condition of insurance coverage, rates, or benefits. Unfortunately, this legislation defines "genetic test" narrowly, to mean "a test using deoxyribonucleic acid extracted from an individual's cells in order to determine the presence of a genetic disease or disorder or the individual's predisposition for a particular genetic disease or disorder."[31] Other ways to "determine the presence of a genetic disease . . . " are currently used to test for PKU, sickle-cell anemia, and other conditions, and it is not clear that they are covered by this legislation.

We need strong laws at the federal level to control genetic discrimination in employment and insurance. Scientists involved in predictive genetics and the Human Genome Project have promised that genetic predictions will improve preventive measures and so make us healthier. However, if insurance companies can use results of genetic tests to limit or deny coverage, such predictions will have the opposite effect. Without coverage, people will have less access to preventive care, thus will be more likely to become ill and less able to get appropriate medical treatments.

Since much of the scientific research that can lead to genetic discrimination is being done in this country, Americans have a special responsibility to develop ways to counteract this insidious new form of discrimination. I hope that scientists will join in the effort, and devote as much energy to preventing genetic discrimination as they do to developing the technologies that make such discrimination possible.

. . . .

ELEVEN

DNA-BASED IDENTIFICATION SYSTEMS, PRIVACY, AND CIVIL LIBERTIES

DNA AND THE CRIMINAL JUSTICE SYSTEM

As reported crime rates soar, so does public discontent with the way the police, investigative agencies, and the courts are handling the apprehension and prosecution of presumed criminals. So, just as the medical system is turning to DNA for a quick fix for people's health problems and the schools are using it to explain children's failure to learn, law enforcement agencies are looking to it for an answer to crime. Earlier, I wrote of the attempt to use the false correlation of aggression with the XYY chromosome condition to explain "criminality." More recently, law enforcement officials have begun looking at DNA in the hope of developing a system for the positive identification of criminals.

The logic is straightforward enough. Except for the rare individual who has an identical twin, each of us is genetically unique. My DNA is different from that of anyone else and, if it were possible to identify the complete base sequence of my DNA, it could be used to identify me absolutely. Since it is not yet possible to do that, or even to identify individual characteristics of it unequivocally, scientists have come up with approximations—techniques that promise to identify a person to a probability of, say, one in a million or, better yet, a billion.

With such a technique in hand, forensic scientists need only a small sample of tissue—a hair or a spot of dried blood or semen—that a perpetrator has left at the site of a crime, or a victim may have left on the body or clothing of a suspect. They can then compare the DNA-profile of that sample with the profile of a blood sample taken from a suspected perpetrator, or from the victim and, ideally, can get a decisive match or an exclusion. In theory, DNA-based profiles are absolute identifiers, like fingerprints, only less subject to deterioration or tampering and more likely to be retrieved as evidence. Advocates of this new procedure call it *DNA-fingerprinting,* though I will avoid that term because at present DNA-based identifications are not nearly as unequivocal as fingerprints can be.

Like fingerprints, forensic DNA samples can be degraded or contaminated. If, for example, the sample has been collected from clothing or a rug that was recently cleaned with a synthetic detergent, residues from the detergent can change the DNA so that restriction enzymes will cut it differently than they would have cut the original sample. Also, tissue or blood samples can easily be contaminated by bacteria, in which case the bacterial DNA becomes part of the sample and will produce misleading results. Both these problems would lead to false exclusions, while other problems can lead to false matches.

In one criminal prosecution, a man named José Castro was indicted for the murder of a neighbor and her two-year-old daughter in the Bronx. The prosecution claimed that a spot of blood that was found on Castro's watch had been identified by DNA typing as coming from the murdered woman.[1] Lifecodes Corporation of Valhalla, New York, the commercial firm that performed the DNA match, asserted that the DNA pattern of the sample on the watch matched that of the murdered woman and that the odds of finding that pattern in the general population were 1:189,200,000, which made the identification sound pretty decisive.

When the case came into court in early 1989, two lawyers, Peter Neufeld and Barry Scheck, decided to challenge this evidence. They called in Eric Lander, a geneticist and mathematician at the Whitehead Institute in Cambridge, Massachusetts, as an expert witness. Lander was disturbed by the poor scientific quality of the data Lifecodes presented to establish the supposed match. He estimated the odds for a random match, which Lifecodes had said were around one in two hundred million, as one in twenty-four—not a very convincing identification.[2] Then something unprecedented happened: Expert witnesses for both the prosecution and defense met together in New York and after evaluating all the data, they issued a consensus statement in which they challenged the adequacy of the evidence Lifecodes had presented as the basis for its statement that the DNA in the two samples was the same. As a result, the judge disallowed the use of this evidence. (Despite this, and to the dismay of Neufeld and Scheck, Castro pleaded guilty and was sentenced to a lengthy prison term.)

The failure of DNA identification in the Castro case has called into question evidence both Lifecodes and Cellmark Diagnostics, the other major DNA-matching firm in the United States, have given in previous cases. Since the Castro case, DNA-profiles have been disallowed in many state courts, though they have been admitted into evidence in others.

To understand this controversy, we need to look more closely at how DNA matching is done. As we saw in chapter 4, identifiable fragments of DNA, called RFLPs, can be produced by letting restriction enzymes chop up DNA into pieces of different lengths. For purposes of genetic diagnosis, this process can then be used to differentiate between family members who carry a particular allele and those who do not.

The technique used to generate DNA-profiles for use in forensics is similar, except that here scientists are not looking for genes. The method is based on the fact that, for some unknown reason, occasional short sequences of base pairs will be repeated over and over, sometimes more than a hundred times. These repeats lie next to one another (*in tandem*), and our chromosomes all contain stretches of DNA made up of such repeating sequences. Because the chromosomes of different people vary in the number of these repeats, such sequences are called *variable number of tandem repeats, or VNTRs*. VNTRs appear to be randomly interspersed in the human genome, and are not known to have any biological function. Forensic scientists have settled on analyzing three or four specific VNTRs for purposes of DNA-based identification.

As with other kinds of RFLPs, VNTRs are identified by digesting a sample of DNA with a battery of restriction enzymes. The mix is placed at the top of a gel on which VNTRs of different size travel at different rates. The different VNTRs are then made visible by tagging them with specific radioactive markers. In theory, since different people's VNTRs are of different lengths, they will move at different rates on the gel, so that each person should have his or her own characteristic pattern.

What makes this method particularly attractive for purposes of criminal investigation is that VNTR analysis requires very little material, so that traces of tissue left at the site of a crime are often enough. If they aren't, a technique called the *polymerase chain reaction, or PCR,* makes it possible to copy minute samples of DNA over and over many times, so that a few molecules of DNA are sufficient to generate a sample large enough for analysis.

The simplicity of the technique and the supposed decisiveness of the identifications it provides have led to attempts to introduce DNA matching in over a hundred recent litigations at the state level, as well as in a few federal cases. It is also being used in Canada, Great Britain, and on the European continent. The FBI has set up its own laboratory for DNA typing at Quantico, Virginia, and private firms such as Lifecodes and Cellmark have set up commercial laboratories. The U.S. Department of

Justice is supporting university-based research into VNTR matching and other potential methods of DNA profiling.

Scientific Problems with DNA Profiles

So what went wrong in the Castro case and what are some of the issues that require a more critical look?

There are both technical and theoretical problems with VNTR identifications. For one thing, mistakes can happen and samples can get mixed up, as they apparently did in another Bronx case in 1987.[3] In the Castro case, when Lifecodes compared the gels on which its scientists had run the DNA from the blood sample found on Castro's watch and the DNA isolated from the murdered woman, they decided to ignore the fact that the positions of the VNTR bands did not match up precisely. They rationalized these differences by saying that the gels were just a little different, but they ran no controls to check whether this was in fact the case. It was on this basis that Lander disputed their evidence and the scientific experts on both sides ended up discounting it.

In other words, forensic laboratories sometimes fail to use accepted scientific standards of what constitutes a match. That seems easy enough to repair. But even when used properly, the matching technique is not all that good. When the FBI's Forensic Laboratory ran profiles on samples of DNA prepared from blood drawn from 225 FBI agents, and the tests were repeated a second time with the same samples and by the same laboratory, one in six of the results did not match up.[4]

In another test of the procedures, the California Association of Crime Laboratories sent fifty samples each to Lifecodes, Cellmark, and Cetus Corporation, a biotechnology firm in Emeryville, California, which performs DNA-based identifications by a somewhat different technique than the other two firms. Both Cetus and Cellmark mistakenly matched samples that were not identical. Lifecodes got all the matches right, but it turned out that the test samples had been run by their research scientists rather than by the technicians who normally perform DNA-matches.[5]

There is another technical problem. Let us imagine that a specific VNTR in my DNA consists of 112 repeats, but in my neighbor's DNA the same VNTR has only 109 repeats. That is a genuine difference between us, but the available techniques are not sensitive enough to pick up such small variations and will report our two VNTRs as identical. When technicians compare VNTRs, the larger the number of repeats in a given VNTR, the greater the difference in the number of repeats between two people has to be in order for that difference to be detected.

Even if the matching techniques become more precise, there is a more fundamental and decisive question. Against what reference population is one to judge the probability that a "randomly chosen" individual will not have the same pattern of VNTRs as someone else? What constitutes a genetically "random" population? Two geneticists, Richard Lewontin and Daniel Hartl, have discussed this issue in *Science* magazine.[6]

For identification purposes, the FBI has established reference populations, which they call "Caucasian," "Black," or "Hispanic." Each group is assumed to be homogeneous and people are assumed always to select their mates at random from within their own group. So, by averaging a few samples from each of these populations, reference samples have been constructed that serve as the standard VNTR pattern from which to make statements about the probability of finding a match within that population.

All these assumptions are problematic, but let us just tackle two obvious ones: that the populations are homogeneous and that there is random mating within each. The U.S. "Caucasian" population consists of immigrants from all over Europe, some of whom have arrived recently, some many generations ago. Some have come from small villages where their ancestors lived for hundreds of years, others from large, cosmopolitan cities. In this country, "Caucasians" often live and marry within fairly distinct communities of Italian-Americans, Swedish-Americans, Irish-Americans, and so on. Yet the statement that the odds are less than one in a hundred million that two individuals have the same pattern of VNTRs is based on the assumption of random sampling from a fictitious, homogeneous, and randomly interbreeding "Caucasian" community.

A similar argument can be made about the census category "Black." It may be convenient, but it has no genealogical or biological meaning. U.S.-born African Americans may come from small rural communities in the South, where their ancestors have lived since they were brought to this continent, or from families who moved north and live in industrial centers like Chicago or Detroit. Some "Blacks" have immigrated recently from Barbados or Jamaica, or from one of the African states. How similar genealogically is a "Black" who has recently immigrated from Trinidad, a second- or third-generation "Black" from Harlem, and a "Black" farmer from Mississippi?

The category "Hispanic" is even less meaningful, since the term includes Caucasians from the United States, Central and South America, and the Caribbean, Cuban and Puerto Rican Blacks, and Native Americans from all over Latin America.

The United States has never been, and is not now, a genealogical melting pot. Any model built on the existence of a well-homogenized, randomly mating population is a fiction and bound to fail. An extreme example of this came in *Texas* v. *Hicks,* in which a man was sentenced to death for a murder which he staunchly denied having committed. DNA-matching was said to show he was the murderer, and the testing laboratory quoted the usual astronomical odds against another person's DNA matching his closely enough to provide a false result. However, as Eric Lander points out, "The crime occurred in a small, inbred town founded by a handful of families," so the probability of finding another person with the same DNA-profile (within the reliability of the technique) must have been considerably greater than was reported.[7]

The scientific community itself is divided over the validity of DNA matching. When Lewontin and Hartl submitted their critique of the technique to *Science,* its editor asked Ranajit Chakraborty and Kenneth Kidd, two scientists known to support the technique, to write a rebuttal for the same issue.[8] As is often the case, this is not a purely scientific disagreement. A news article in *Nature* revealed that Chakraborty is a coinvestigator on a $300,000 grant for DNA forensics research from the National Institute of Justice, which is funded by the U.S. Department of Justice.[9] The Department of Justice is by no means a disinterested party in this discussion. It is deeply committed to the use of "DNA fingerprinting," and both *Science* and *Nature* reported attempts by a Department official to derail the publication of Lewontin and Hartl's article.[10] The Justice Department's intervention in this supposedly scientific debate is both inappropriate and frightening. If the value of the technique is still in question, surely it is in the interests of the criminal justice system to resolve the debate rather than to bury it.

Since the publication of the opposing articles, several scientists have written letters to *Science* on both sides, and the original authors have responded in turn.[11] Clearly, disagreement is not dying down. The situation has become so confusing that when a National Academy of Sciences panel in April 1992 issued what was supposed to be the authoritative statement on the value of DNA-based forensic techniques,[12] this report was interpreted in diametrically opposite ways. The *New York Times* ran an article about it under the headline, "U.S. Panel Seeking Restriction on Use of DNA in Courts," adding the subheading "Judges Are Asked to Bar Genetic 'Fingerprinting' Until Basis in Science Is Stronger."[13] Then, the very next day, the *Times* and other papers ran an Associated Press dispatch which explicitly contradicted the first article.[14] The AP story quotes Victor McKusick, the chairman of the National Academy panel, as saying, "We think [genetic fingerprinting] is a powerful tool for criminal investigation and for exoneration of innocent individuals and one that should be used even as standards are strengthened." McKusick said that the first *Times* article "seriously misrepresents our findings," but *Times* writer Gina Kolata stands by her story, and my own reading of the report bears her out.

As usual, there was also a question of conflict of interest. Several members of the National Academy panel serve on boards of directors or have financial interests both in genetic screening companies and in companies involved in "DNA-fingerprinting."[15] One such member, C. Thomas Caskey, resigned from this panel in 1991 after a *Nature* article revealed his financial links to a company doing DNA-based identifications.[16]

What all this boils down to is that, at present, the reliability of the data and of the scientists producing those data are in doubt. Unfortunately, juries and judges, like the rest of the public, are easily swayed by the mystique and power of science. When it is offered with appropriate fanfare, a DNA match need not be scientifically reliable to prove decisive in a court of law.

Apparently, our elected representatives in the House and the Senate are equally impressed by the mystique of DNA matching. Despite all the problems with DNA-based identification, Senator Paul Simon of Illinois and Representative Don Edwards of California (both Democrats) have introduced legislation to authorize the FBI to set standards for DNA testing. This move was sufficiently startling that *Science* magazine titled its report about this legislation, "Letting the 'Cops' make the Rules for DNA Fingerprints."[17]

Such a move is worrisome not only for the reasons this title suggests. In the past, the FBI has consistently obstructed the introduction of regulations and standards for DNA matching. It has opposed independent testing of its own results, as well as proposals to require laboratories to document their conclusions in writing and to have the scientists and technicians who perform the tests sign their reports. As a result, at present, "no private or public crime laboratory [in this country] . . . is regulated by any government agency," so that "there is more regulation of clinical laboratories that determine whether one has mononucleosis than there is of forensic laboratories able to produce DNA test results that can help send a person to the electric chair."[18]

Despite the problems and disagreements, law enforcement officials in many areas are going ahead and using DNA technologies, and are beginning to store tissue samples and set up data banks of "DNA fingerprints." In an article published in *Parade Magazine* in March 1991, Earl Ubell writes that "several states now are taking blood samples from convicted rapists and other violent criminals. Their DNA profiles will be stored in a data bank for use by police across the United States." In support of this practice, he writes: "Using DNA fingerprinting, for example, detectives could trace a rapist convicted in Utah who later rapes in Ohio by matching the DNA 'prints' on file with those in traces found on the victims." [19]

So, DNA-based identification, though highly questionable in its present form, is being sold to a terrified public as a way to solve the heinous crimes we hear about every day. This is a quick fix, rather than a real solution. Blaming yet another crime on a convicted rapist or murderer may make law enforcement officers look (and feel) better but, if the charge is built on unreliable evidence, it does not make us any safer on the street or in our homes.

There is another important issue. We are regularly told that DNA profiles will be at least as useful to defendants as they are to prosecutors, since they are as likely to establish innocence as guilt. However, unless the technology becomes cheaper and more accessible than it is now, its acceptance by the courts disproportionately increases the advantage of the prosecution. Defendants and their attorneys usually do not have the funds to employ this sort of technology.

GENETIC PRIVACY AND CIVIL LIBERTIES

At present, efforts are under way to develop DNA profiles that will identify a specific individual unequivocally, without drawing on his or her relationship to a reference population. When such profiles become feasible DNA analysis will indeed provide individual "fingerprints." While this may clear up some problems, a whole range of privacy and civil liberties issues still need to be considered.

Once DNA has been isolated from a sample of blood or some other tissue, it can be used for other purposes than the one for which it was obtained. Since DNA can yield information about a person's health and about matters of social import, such as paternity, the collection of DNA samples always constitutes a potential invasion of privacy. This is so even when the analyses may not be as informative as their proponents claim they are. If people believe that untold quantities of significant information are coded into our DNA, and that scientists are daily becoming more able to decipher it, this can pressure them into giving up their right to privacy and divulging long-standing family secrets.

To prevent such misuses and pressures, it must be made legally impossible for any person or group to collect samples for DNA typing from people without their informed consent. Where such consent can be legally circumvented, as for example in criminal investigations, it is essential to have regulations to ensure that samples can be used only for the purpose for which they were collected and that they will be destroyed once that purpose has been fulfilled.

This is not what is envisaged by law enforcement officials, nor is it written into the Human Genome Privacy Act, the only federal legislation that has so far been drafted to regulate the collection, dissemination, and storage of DNA-based information. Even though this legislation may not be enacted in its present form, the inadequacy of its proposed protections of privacy and civil liberties points up the need for vigilance.

In ordinary practice, a court order usually suffices to permit police or other officials to collect evidence by means that would otherwise be considered invasive, such as drawing blood or obtaining seminal fluid. While the preservation of all evidentiary material for the entire duration of a legal process could potentially benefit both the prosecution and the defendant, the question is how to prevent its being stored after that.

There are no consistent rules about the banking of tissues or body fluids collected from suspects, defendants, or persons convicted of a felony. The Human Genome Privacy Act uses language developed with reference to credit transactions. The stated purpose of the Act is

> to safeguard individual privacy of genetic information from the misuse of records maintained by [federal] agencies or their contractors or grantees for the purpose of research, diagnosis, treatment, or identification of genetic disorders, and to provide to individuals access to records concerning their genome which are maintained by agencies for any purpose.[20]

However, the bill gives access to such information too readily to keep people's genetic or personal health information private. Also the bill is built on the premise that it is acceptable for private and government bodies to collect, store, and disperse genetic information as long as they do not "misuse" that information. The Act does not address the problems that are inherent in collecting this information even for its "intended" use.

In January 1992, the Defense Department announced plans to "establish a repository of genetic information on all American service members as a new way of identifying future casualties of war."[21] While the Defense Department hails this as a way to avoid any more "unknown soldiers," it could pave the way for other government and private organizations to establish DNA data banks. The military is one of the largest employers in the United States. Allowing them to implement such a program is no small matter.

The collection and dissemination of samples for DNA-based identification, as well as the storage of such information, needs to be outlawed except under specific, narrowly defined circumstances. In the rare event that such activities are permitted, they must be closely regulated and supervised. For example, the storage of such samples or information, after they have served the specific purpose for which they were obtained, must be prohibited; the samples or information must not be shared by different agencies or used for other purposes than the one for which they were collected; people must be guaranteed access to their own records; and, where errors are uncovered, they must be corrected and the mistaken information expunged.

EMPLOYMENT- AND HEALTH-RELATED ISSUES

As we saw in the previous chapter, employers are within the law when they invade the genetic privacy of employees by performing tests that are said to predict the employees' risk of developing employment-related diseases. The Americans with Disabilities Act can be interpreted to prohibit comparable preemployment screening, but employers can easily get around that by obtaining the job applicant's consent. Where obtaining a job depends on a person's giving this "consent," the word loses its meaning.

As I write, I am looking at a copy of a "Release of Information" form that applicants for U.S. government jobs at even the lowest levels are asked to sign "voluntarily." It authorizes "any duly accredited representative of the Federal Government, including those from . . . the Federal Bureau of Investigation . . . to obtain any information relating to [the applicant's] activities from schools, residential management agents, employers, . . . *medical institutions, hospitals or other repositories of medical records,* or individuals. This information may include, but is not limited to, [the applicant's] academic [record], . . . personal history, . . . *medical, psychiatric/psychological* . . . information"[22] (italics mine).

The release authorizes any individual to give up such information "upon request of the duly accredited representative of any authorized agency regardless of any agreement [the applicant] may have made with [that individual] previously to the contrary" and specifies that the "users [of this information] may redisclose [it] as authorized by law." The fact that all job applicants are told to "read this authorization, . . . then sign and date it in ink," makes a mockery of the notion that the consent is voluntary.

Moving on to a medical context, how can we guarantee the genetic privacy of the "healthy ill"? What are the obligations of a physician to inform a person's close relatives, such as parents, siblings, children, or a present or future marriage partner, if that person is diagnosed to have a genetic condition or "predisposition," but has no symptoms and perhaps will never develop any? Does the state have the right to force a physician or the individual to whom the genetic diagnosis applies to inform close relatives or a future spouse? A genetic condition is not contagious, like some bacterial or viral diseases, but it may be transmitted to one's children, and in certain cases (such as PKU) preventative therapies may be available.

France has a privacy law that prevents physicians from divulging genetic information to anyone, including the individual to whom the information refers, unless it was collected at his or her request. This law has come to public notice because a group of French researchers have identified a large number of people who may carry an allele implicated in an inherited form of juvenile glaucoma, an eye disease that can lead to blindness. These people were not identified through genetic tests, but rather by tracing their descent from a couple who lived in northwestern France in the fifteenth century.

While physicians have been alerted to the fact that people in their area are carriers of this allele, they cannot be given names of specific people, nor can the carriers themselves be informed. This is because, under French law, "distributing a list of individuals obtained by a genealogic study would constitute an authoritarian public health measure that would infringe on individual liberty and privacy. . . . Circulating the names of potential carriers of genes predisposing to diseases might lead to discrimination in hiring or insurance."[23] When the law bars people from getting information about themselves, it may be going too far but, as we saw in the last chapter, unless such information is kept strictly confidential, it is all too easy for it to be used in discriminatory ways.

In the United States, patients' hospital records are accessible to both researchers and hospital staff. This includes everyone from the treating physician to the pharmacist and to quality assurance personnel and billing clerks. Although the patients' names are not to be noted or cited unless this information is needed for treatment or billing, they are part of the medical record. When it comes to the information health insurers keep on policyholders, such information gets shared among insurance companies. As a result, many people have found that when they move from one part of the country to another and change their health insurer, the new company knows things about them that they did not include in their new application. Insurance information is banked and health insurers have access to it. There is every reason to assume that genetic test results will make up an increasingly significant portion of that information, unless laws are enacted to prohibit this.

The potential for infringing on people's privacy is enormous because, whether a specific piece of genetic information is meaningful or not and whether or not it is accurate, all genetic information projects into the future. It has implications for the assessment not only of one person's health but for the health of all of that person's relatives, descendants, or potential descendants.

CONTROLLING GENETIC INFORMATION

When Congress passed the Social Security Act in 1934, it provided that, in the interest of confidentiality and privacy, Social Security numbers should be used only with reference to the Social Security system. Today our Social Security number is a tag that identifies us on state and federal income tax returns and on our driver's licenses in many states, unless we request otherwise. We are sometimes asked for it when we try to cash a check, as well as in numerous other contexts that have nothing to do with Social Security. It has become our identifying label in hundreds of computerized information banks. The same is true of fingerprints. They were introduced as a way to identify suspected criminals, but many of us have been fingerprinted numerous times for employment and for professional licenses, although we never have been involved in a criminal proceeding.

If we permit DNA-profiles to become a tool of law-enforcement, we can be sure that the information will be used in other ways as well. As I said before, DNA-profiles and the blood or tissue samples from which they are drawn can be made to yield information about a wide range of matters. If we allow the information or, worse yet, the samples to be stored in computerized information banks, they are bound to be used for other purposes than just the one(s) for which they were obtained in the first place.

Proponents of DNA-typing and data-banking may emphasize the technologies' potential usefulness as tools for tracking serial rapists and murderers, identifying war casualties, or finding missing children or amnesiac grandparents, but this is just a way to gather public support for these activities. Past history gives us no reason to trust assurances by the FBI, other law enforcement agencies, or the armed services that such information will not be used in other contexts. Government agencies want to have the means to collect as much information as they can about each of us. It is up to us to prevent such encroachments on our privacy and our constitutionally guaranteed liberties.

There is no reason to believe that collecting DNA-profiles and assembling them into data bases will benefit society. Most violent crimes are not committed by repeat-offenders, so that storing samples for future identification is neither necessary nor cost-effective. Furthermore, such practices clearly endanger our fragile guarantees of civil liberties and privacy.

Some groups are beginning to mobilize against the premature or ill-considered use of the new technologies. When the question of taking and storing samples for DNA-typing came up in King County, Washington, the citizens' committee established to look into the proposal pointed out that the proponents had not demonstrated what the benefits of this expensive new procedure would be. The committee urged that research first be conducted to evaluate the claims that this practice would help solve crimes or convict defendants.[24]

So far, the results of forensic genetic matching have been thoroughly equivocal. Colin Pitchfork, a British rapist-murderer, is said to be the first criminal to have been identified on the basis of DNA-typing. In fact, he came to the attention of the police not because of his DNA-profile, but because of the myth of its infallibility. Pitchfork became a suspect only because he was so frightened by the reputation of DNA-typing that he got a friend secretly to give blood in his place in order to avoid participating in a dragnet in which the police collected blood samples from 5,512 men who lived or worked in the neighborhood where the crimes had been committed. His deception led the police to track him down and we will never know whether he would have been identified had he provided a blood sample along with the other men in his village.[25] But all the coverage of this case focused on the success of the procedure. The fact that the police pressured all those men to submit "voluntary" blood samples received little notice. The attention went to the supposed power of the technique, not to its coercive application.

Let us remember that the mystique of scientific and technical progress lends a power to DNA-typing and banking that the technology may not merit. Whether or not the results are scientifically accurate, the "evidence" can be used to sway juries and judges and to intimidate citizens into renouncing their constitutional rights in various contexts. Without vigilance on our part, in the near future Big Brother could be collecting and computerizing our DNA-profiles and then disseminating the results, without either our knowledge or our consent.

TWELVE

· ·

IN CONCLUSION . . .

As I have suggested throughout this book, interesting research is being done in molecular biology, but there are also many dangers. If I have concentrated on the dangers it is because they are so frequently ignored, downplayed, or flatly denied.

Research will be able to answer questions about the diversity within different groups of organisms, including human populations, and about the evolutionary relationships among them. Increased knowledge about DNA also will improve our understanding of the structure and function of different proteins. It is less likely to improve our understanding of the network of metabolic relationships underlying most diseases and disabilities and the complex processes of growth and development, because these depend on many factors that are not influenced by genes.

And yet, DNA is frequently discussed as though it were the be-all and end-all of biology. A recent promotional video for the Genome Project starts with an announcer comparing the Project's work to the voyages of the European renaissance explorers. "Imagine a map that would lead us to the richest treasure in the world," the video begins. "Not a treasure of jewels or gold, but a treasure far more important to humankind. This treasure is knowledge, the ability to chart our genetic blueprint. We haven't deciphered that map yet, but when the Human Genome Project is completed we will know exactly where in the cells of our bodies every genetic inheritance of humankind is to be found."[1]

The "map" metaphor has long been a favorite of geneticists. The problem is that while geographic maps have obvious uses, it is not clear what one could do with a map of the entire human genome if one had it. While sequencing certain sections of the genome may be useful to scien-

158

tists, mapping the entire genome will not tell us "exactly where . . . every genetic inheritance of humankind is to be found." Indeed, the phrase is meaningless, though it conveys endless promise.

Not surprisingly, such grand promises make up the better part of the video. Toward the end, however, the video turns to the social and ethical consequences of the Genome Project. James Watson, the Project's first director, speaks of the problem of genetic discrimination, and of the need to protect people affected by an "unjust throw of the genetic dice." Dr. Nancy Wexler, a psychologist at Columbia University, who chairs the Ethical, Legal, and Social Implications study group (ELSI) of the Genome Project, sketches that group's goals.

It is encouraging that the Genome Project has such a study group, and that the scientists are concerned about the deleterious implications of their research. However, the ELSI program will not affect decisions about what scientific work is done by the Genome Project. It can only make suggestions about public policy questions, such as how to protect people from genetic discrimination and invasions of privacy. It has no power to insure that these suggestions translate into policies. Should ELSI recommend against certain sorts of experiments or genetic tests because studies it has sponsored suggest that these will have dangerous consequences, many of the big names in the Genome Project are likely to be arrayed on the other side.

The ELSI program gives the illusion that the social problems raised by the Genome Project will be handled, but the way it has been set up virtually guarantees that such concerns will not get in the way of the science. At some time in the future, economic, social, and ethical issues may force genome scientists to modify their projects or even terminate one or another of them. However, it is unlikely that the ELSI program will force such changes.

While ELSI will have no power to shape policy, it is already affecting the debate on ethical questions about molecular biological research. Though the Genome Project has allocated only 5 percent of its budget to this program, that is still a great deal of money for studies on the social and ethical impact of science and technology. As a result, more and more potential critics of the Genome Project are obtaining their funding from the Project. This puts the Genome Project in a position to supervise what questions are asked and to define the parameters of the debate.

There is another aspect to consider. Whatever conclusions researchers funded by ELSI may come to, and however much those conclusions may be respected by scientists, the fact remains that scientists will not, in the end, have the power to control the ways in which the results of their research are used. This is illustrated by the history of the development and use of the atomic bomb and later generations of nuclear weapons. The scientists in the Manhattan Project were by no means agreed that atom bombs should be dropped on Japan. Many suggested less destructive alternatives, such as making a public demonstration of the bomb's power. Many were horrified when the bombs were dropped; some went on to be leaders in the antinuclear movement, and a number left physics altogether.

Nuclear research, like genetic research, is not all good or all bad. Radioactive isotopes have been useful in nuclear medicine, for both diagnosis and treatment. Even nuclear power might be made safe under ideal circumstances. The problem is that circumstances are far from ideal. In a world where cutting costs and maximizing profits are the order of the day, safety is not a top priority. Corners are cut, risks are overlooked, hasty decisions are made, and major problems like nuclear waste disposal are simply swept under the carpet.

Imagine that genome scientists and human geneticists were to call for a moratorium on the use of genetic tests for decisions affecting employment or insurance until legislation had been enacted to protect people against genetic discrimination. In the first place, it is unlikely that they would all be of one mind about this. But even if they were, it is questionable whether they would have the power to bring such a moratorium about. Furthermore, antidiscrimination laws would not solve the problem. Important as civil rights laws have been, they have not stopped other sorts of discrimination. Often, the changes have been largely cosmetic. The Civil Rights Act of 1964 outlaws discrimination on the basis of race or sex. Nevertheless, twelve years later in *General Electric Company* v. *Gilbert,* a majority of the Supreme Court decided that a company insurance plan that excluded pregnancy-related disabilities did not discriminate against women. As the dissenting minority pointed out, in effect the majority ruled that the plan did not contravene civil rights law because men as well as women were excluded from pregnancy coverage and women, like men, were covered for prostate surgery.[2]

Anyway, all of this talk about whether scientists will or will not decide to do the right thing is really beside the point. The issues are too big, and affect all of us too deeply, to be left up to scientists or other sorts of experts. Geneticists, whatever their political, ethical, or social views, are interested in genetics and wish to see their field move forward. They should not be the ones to decide matters of policy about what should and should not be done. If you want to build a skyscraper, you need an architect who specializes in building skyscrapers, but if you want a panel to decide whether or not to build more skyscrapers, you do not want it to consist mainly of those architects. Nor, in the case of either genetics or skyscrapers, should the decisions be based on the interests of businesses which stand to profit from them.

I do not mean to suggest that we need more laws forbidding certain kinds of genetic research, though I can imagine technologies that should be outlawed because they constitute threats to health, privacy, or civil rights. What we need is wider public debate about the ways in which genetic research and technologies are likely to affect us. Virtually all scientific research in this country is done at least in part at public expense. Therefore, the public should be involved in the decisions about how research funds are allocated.

We also need to make sure that new technologies do not create new limitations. Genetic testing, like HIV testing, should never be compulsory or even routine. While tests should be available to those who want them, there needs to be careful counseling so that people understand the implications of all possible test results before they have the tests. They need to understand the implications of genetic testing not only for their health, but also for their relationships with friends and family. They also need to consider how the results—and even the fact that they elect to have such tests—could affect their prospects for employment and their insurability.

Beyond asking what genetic research should be done and how it should be applied, we need to question the current emphasis on genes as determining our development, health, and behavior. Focusing on genes leads almost inevitably to an assignment of values: these genes are good, those genes are bad. We may start with relatively clear-cut cases like Tay-Sachs disease, which is invariably fatal in early childhood, but we almost immediately get into gray areas where people leading quite ordinary lives can suddenly find themselves stigmatized as defective.

Scientists and physicians should not be given the right to assign such labels, but the problem is greater than that. The labels themselves are inherently wrong, no matter who is doing the labeling. There is no way to say which lives are or are not valuable. I am glad Woody Guthrie was born, though he developed Huntington disease. I am glad for all the blind poets and musicians, from Homer to Stevie Wonder. Who knows, maybe Helen Keller would have led a completely undistinguished life instead of becoming a famous writer and political activist had her immune system not failed her as a child.

All of us are flawed according to someone's standards, and will continue to be, no matter what scientific breakthroughs may come along. No one, and no group of people have, in the words of Hannah Arendt, "any right to determine who should and who should not inhabit the world."[3]

Not only do we need to be careful about the power the genetic ideology gives to scientific and medical experts, lawmakers, and society, we also need to be aware of how this ideology can affect our thinking about ourselves. We urgently need to demedicalize our relationship to our bodies and our state of health. At present, babies enter the world preceded by ultrasound pictures and genetic predictions. That entry itself is a medical, and all too often a surgical, event. We live from checkup to checkup, test to test, injection to injection, pill to pill. All too often, we die amid a web of tubes and wires, our exit heralded by ringing bells and flashing lights. That the healthy as well as the ill live under such continuous medical surveillance is in the interest of the medical-industrial complex, and not in ours.

Our new fixation on genes can only make us less confident about our bodily functioning and so increase our alienation from ourselves. We need to engage in active debates about the practical consequences of genetic forecasts for our self-image, our health, our work lives, our social relationships, and our privacy.

To be ill or disabled is part of the human condition, and not the worst thing that can happen to us. Far worse to harden ourselves and look on people who are ill or have disabilities as statistics or as burdens, to be prevented at all costs. In the name of disease prevention, the genetic ideology in the past led to gross abuses of power. We must see to it that the new technical knowledge does not outstrip our political capabilities to show all our fellow humans the respect and good will we would have them show us.

APPENDIX

. . . .

MITOCHONDRIAL DNA

MITOCHONDRIA AND THEIR DNA

In this book, we have only looked at the DNA in the nucleus of a cell, the *nuclear DNA*. But the cell substance, or *cytoplasm*, contains numerous microscopic and submicroscopic particles in addition to a nucleus. One kind of cytoplasmic particles in the cell, which are called *mitochondria*, have their own separate complement of DNA.

Mitochondrial and nuclear DNA differ in a number of ways. While human nuclear DNA is organized into strands that form twenty-three pairs of chromosomes, the DNA in each mitochondrion forms a single, continuous, circular chromosome. Also, our cell nuclei contain much more DNA than our mitochondria do. Whereas three billion base pairs are distributed among the human nuclear chromosomes, the human mitochondrial chromosome contains only 16,569 base pairs.

However, most of our cells contain only one nucleus (striated muscle cells, which contain many nuclei, are an exception), while cells in different tissues contain anywhere from a few to several thousand mitochondria, and mammalian eggs contain about ten thousand. Scientists estimate that we each have a total of some 10^{16} (that is, a one with sixteen zeros after it) mitochondria in our cells and, except for a few random mutations, the base sequence of the mitochondrial, as well as the nuclear, DNA is the same in all of an individual's cells.

Since the mitochondrial genome is much smaller than the nuclear genome, molecular biologists have already established the base sequence of mitochondrial DNA in humans as well as in several other animals. Interestingly enough, all these mitochondrial genomes are of pretty much the same size and are involved in the synthesis of similar products. (However, the mitochondrial genome of yeast is much larger than ours.)

Mitochondria function in the oxidation of foodstuffs and convert the energy contained in the foods we eat into a form that the various cells and tissues in our bodies can use to carry out their appropriate functions. Mitochondrial genes are implicated in the synthesis of some, though by no means all, the enzymes and other proteins contained in the mitochondria, and nuclear genes also participate in the synthesis of some essential components of mitochondria.

A widely accepted theory, first proposed in 1918 by a French biologist named Paul Poitier, holds that mitochondria are derived from small bacteria that were incorporated by larger single-celled organisms hundreds of millions of years ago.[1] This theory assumes that over the ages these two kinds of organisms established a mutually beneficial symbiotic relationship and that eventually the bacteria lost the ability to exist on their own and evolved into indispensable parts of the larger cells. Later research has supported this theory, showing that the DNA in the circular chromosome of mitochondria is organized very much like the DNA in present-day bacteria.

When a cell divides, the mitochondria in its cytoplasm distribute themselves between the two daughter cells. Just as cells and their nuclei divide, so do mitochondria, except that they divide on their own schedule, which is not synchronous with cellular and nuclear division. The mitochondria first grow larger, then pinch off daughter mitochondria in a way that resembles the reproductive process of bacteria.

Scientists have only lately begun to explore the possibility that mitochondrial DNA might play a role in the genesis and transmission of inherited conditions. In the late 1980s a few articles were published that correlated single base changes in mitochondrial DNA with the appearance of conditions that seem to be transmitted only through the maternal line.[2]

It is not clear what proportion of the mitochondria in a cell must contain mutant DNA in order for this type of condition to become noticeable. For example, a recent study attributed a neurological condition in a family to a mutation in mitochondrial DNA. The scientists found that in the three people who exhibited symptoms, over 82 percent of the mitochondria carried a specific mutation, whereas the five family members with less than 34 percent of affected mitochondria showed no symptoms.[3]

Increasing numbers of physicians and molecular biologists are turning their attention to the possibility that mutations in mitochondrial DNA may be implicated in health conditions. They are likely to become aware of new and unanticipated patterns of inheritance that depend on the way different proportions of mitochondria containing various alleles of a gene get sorted out in the genesis of egg cells.

A point to bear in mind is that, whereas our cells contain only two copies of each nuclear gene (one contributed by the mother and the other by the father), cells have as many copies of each mitochondrial gene as they have mitochondria. During cell division, the mitochondria distribute themselves randomly between the two daughter cells. If one of these daughter cells happens to be an egg that subsequently gets fertilized, the new individual that develops from this egg need not have the same representation of different alleles as was present in the cells of the mother.

This might seem to contradict my earlier statement that the base sequence of the mitochondrial DNA is the same in all of an individual's cells, but it doesn't. For all extents and purposes, the mitochondrial DNA is identical. Even though one base in a specific gene may be different and this minute difference may be enough to lead to the manifestation of a condition, this is one out of 16,569 base pairs.

Scientists have suggested that some of the changes people experience as they age may arise from an accumulation of mutations in mitochondrial DNA, which debilitate increasing numbers of mitochondria and decrease their effectiveness in the metabolism of various cells and tissues.[4] Such mutations would not be passed on to a woman's children, but the impairments might become noticeable to the individual in whose tissues they take place.

Molecular biologists have evidence that mitochondrial DNA mutates six to seventeen times as fast as nuclear DNA.[5] There are at least two possible reasons why this is so. One is that enzymes that can repair DNA exist in the nucleus, but not in mitochondria. The other is that since mitochondria, by their nature, metabolize various chemicals that occur in the environment, they may be more subject to environmental hazards than the nucleus is.

The picture that emerges from considerations of the potential contributions of mitochondrial DNA to health and disease has even greater complexity and unpredictability than what we have seen happening with nuclear DNA. We can only hope that this complexity will discourage attempts at simplistic explanations.

INHERITANCE OF MITOCHONDRIAL DNA

There is a big difference in the hereditary transmission of nuclear and mitochondrial DNA because, when a sperm and egg fuse during fertilization, the sperm contributes its nucleus and only minuscule amounts of cytoplasm, hence essentially no mitochondria. The egg, however, contributes not only its nucleus to the embryo but its entire cytoplasm and therefore all its mitochondria and mitochondrial DNA. So, while our nuclear DNA contains equal contributions from both parents, mitochondrial DNA passes undiluted from a mother to all her offspring. This means that each of us is connected to both our parents through our nuclear DNA, but a separate line of inheritance connects us to our mothers and to their mothers before them. A consequence of this is that, whereas parents and their various children all differ in their nuclear DNA, mitochondrial DNA is identical in any woman and her children.

The line of mitochondrial inheritance stops with each son, but is continuous from mothers to their daughters and their daughters' daughters and so on. This continuity has important implications, as it can be used to establish family relationships.

Using the fact that a woman's mitochondrial DNA is essentially identical in all her children, some molecular biologists are developing techniques that could be used to identify missing children and other relatives. Human rights activists hope that such techniques will be useful for tracking children of the disappeared in such countries as Argentina and Chile.

Even when there is no way to recover tissue samples from parents who were disappeared by the government, the mitochondrial DNA of any maternal relative, such as a maternal grandmother or aunt or uncle, is virtually identical to that of a child's biological mother. The mitochondrial DNA from any one of these individuals could therefore be used to link the child and its maternal relatives.[6] This method has an advantage over other available ones because "when there is only one surviving relative and he or she is removed by more than one generation, disputed identity can still be resolved provided that kinship has preserved the maternal lineage."[7]

Scientists are currently testing the reliability of this method by trying it out on people who are known to be related or unrelated in the maternal line. If it proves to be as reliable as is hoped, it could then be used to trace children whose parents were forcibly disappeared and who were given for adoption to members of the ruling elite.

In the late 1980s, a group of molecular biologists at the University of California at Berkeley came up with an ingenious theory about human origins.[8] Based on the similarities and differences in samples of mitochondrial DNA obtained from various geographical populations, they concluded that all present-day human beings (modern *Homo sapiens*) could trace their descent to one woman who lived in Africa some 200,000 years ago. This hypothetical woman came to be known as the "African

Eve." More recently a number of investigators have pointed out both technical problems with the California group's research and conceptual problems with such a recent date for a common human foremother.[9] Though everyone agrees that *hominids* (early human ancestors) originally arose in Africa and spread from there into the rest of the world, it is not at all clear where, when, or how often these hominids evolved into modern humans.

NOTES

1. OF GENES AND PEOPLE

1. Abby Lippman. "Prenatal Genetic Testing and Screening: Constructing Needs and Reinforcing Inequities." *American Journal of Law and Medicine,* vol. 17, 1991, pp. 15–50 (p. 19).

2. Stephen S. Hall. "James Watson and the Search for Biology's 'Holy Grail.'" *Smithsonian,* vol. 20, February 1990, pp. 41–49.

3. Lois Wingerson. *Mapping Our Genes: The Genome Project and the Future of Medicine.* New York: Dutton, 1990, p. 286.

4. Richard Saltus. "Medical Notebook." *Boston Globe,* October 10, 1991, p. 3.

2. GENETIC LABELING AND THE OLD EUGENICS

1. Charles Benedict Davenport. *Heredity in Relation to Eugenics.* New York: Henry Holt and Company, 1913, p. 80.

2. Francis Galton. *Inquiries into Human Faculty.* London: Macmillan, 1883, pp. 24–25.

3. H. J. Muller. "Principles of Heredity." [1912] In H. J. Muller, ed. *Studies in Genetics.* Bloomington: University of Indiana Press, 1962, pp. 6–17.

4. Julian Huxley. "The Vital Importance of Eugenics." *Harper's Monthly,* vol. 163, August 1941, pp. 324–331.

5. Rodger Hurley. *Poverty and Mental Retardation: A Causal Relationship.* New York: Random House, 1969.

6. Lewis M. Terman. "The Conservation of Talent." *School and Society,* vol. 19, 1924, pp. 359–364.

7. Allan Chase. *The Legacy of Malthus: The Social Costs of the New Scientific Racism.* New York: Knopf, 1977, p. 214.

8. Stephan L. Chorover. *From Genesis to Genocide*. Cambridge: MIT Press, 1979.

9. Robert J. Lifton. *The Nazi Doctors*. New York: Basic Books, 1986.

10. Benno Müller-Hill. *Murderous Science*. Oxford: Oxford University Press, 1988.

11. Robert N. Proctor. *Racial Hygiene: Medicine under the Nazis*. Cambridge: Harvard University Press, 1988.

12. Chase. *The Legacy of Malthus*, p. 121.

13. Charles Benedict Davenport. *Heredity in Relation to Eugenics*. New York: Henry Holt and Company, 1913, pp. 80, 82.

14. Chase. *The Legacy of Malthus*, p. 124.

15. Ibid., p. 124.

16. Ibid., p. 125.

17. J. B. S. Haldane. *Heredity and Politics*. New York: W. W. Norton and Co., 1938, pp. 103–104.

18. Ibid., p. 104.

19. Ibid., pp. 104–105.

20. Ibid., p. 105.

21. Phillip R. Reilly. *The Surgical Solution: A History of Involuntary Sterilization in the United States*. Baltimore: Johns Hopkins University Press, 1992, p. 148.

3. THE NEW EUGENICS

1. Helen Rodriguez-Trias. "Sterilization Abuse." In Ruth Hubbard, Mary Sue Henifin, and Barbara Fried, eds. *Biological Woman—The Convenient Myth*. Cambridge, Mass.: Schenkman Publishing Company, 1982, p. 149.

2. Bentley Glass. "Science: Endless Horizons or Golden Age?" *Science*, vol. 171, 1971, pp. 23–29.

3. Joseph F. Fletcher. "Knowledge, Risk, and the Right to Reproduce: A Limiting Principle." In Aubrey Milunsky and George J. Annas, eds. *Genetics and the Law II*. New York: Plenum Press, 1980, pp. 131–135.

4. Margery W. Shaw. "The Potential Plaintiff: Preconception and Prenatal Torts." In Aubrey Milunsky and George J. Annas, eds. *Genetics and the Law II*. New York: Plenum Press, 1980, pp. 225–232.

5. John A. Robertson. "Procreative Liberty and the Control of Conception, Pregnancy, and Childbirth." *Virginia Law Review,* vol. 69, 1983, pp. 405–464 (pp. 438, 450).

6. Daniel J. Kevles. *In the Name of Eugenics: Genetics and the Uses of Human Heredity*. New York: Knopf, 1985, p. 277.

7. Linda Gilkerson. "A Fully Human Life." *Family Resource Coalition Report,* 1988, no. 2, p. 3.

8. Jeffrey R. Botkin and Sonia Alemagno. "Carrier Screening for Cystic Fibrosis: A Pilot Study of the Attitudes of Pregnant Women." *American Journal of Public Health,* vol. 82, 1992, pp. 723–725.

9. Marsha Saxton. "Prenatal Screening and Discriminatory Attitudes About Disability." *Genewatch,* January–February 1987, pp. 8–10.

10. Jean Seligmann with Donna Foote. "Whose Baby Is It, Anyway?" *Newsweek,* October 28, 1991, p. 73.

11. Paul R. Billings, Mel A. Kohn, Margaret de Cuevas, Jonathan Beckwith, Joseph S. Alper, and Marvin Natowicz. "Discrimination as a Consequence of Genetic Testing." *American Journal of Human Genetics,* vol. 50, 1992, pp. 476–482.

12. C. Ezzell. "Gene Discovery May Aid Marfan's Diagnosis." *Science News,* vol. 140, 1991, p. 55.

13. Peter Aldhous. "Who Needs a Genome Ethics Treaty?" *Nature,* vol. 351, 1991, p. 507.

14. Benno Müller-Hill. *Murderous Science.* Oxford: Oxford University Press, 1988; Robert N. Proctor. *Racial Hygiene: Medicine Under the Nazis.* Cambridge: Harvard University Press, 1988.

15. Troy Duster. *Backdoor to Eugenics.* New York: Routledge, 1990, p. 26.

16. Richard A. Knox. "Gene Mutations Found in Cystic Fibrosis." *Boston Globe,* July 24, 1990, p. 6.

17. Benjamin S. Wilfond and Norman Frost. "The Cystic Fibrosis Gene: Medical and Social Implications for Heterozygote Detection." *Journal of the American Medical Association,* vol. 263, 1990, pp. 2777–2783.

18. Ibid., p. 2778.

19. Richard Saltus. "Cystic Fibrosis Test: Is It Really Accurate?" *Boston Globe,* September 12, 1991, p. 3.

20. Troy Duster. *Backdoor to Eugenics.* New York: Routledge, 1990.

21. J. B. S. Haldane. *Heredity and Politics.* New York: W. W. Norton and Co., 1938, pp. 88–89.

22. David T. Suzuki, Anthony J. F. Griffiths, Jeffrey H. Miller, and Richard C. Lewontin. *An Introduction to Genetic Analysis.* (Fourth edition.) New York: W. H. Freeman and Co., 1989, p. 83.

23. Marcia Barinaga. "Novel Function Discovered for the Cystic Fibrosis Gene." *Science,* vol. 256, 1992, pp. 444–445.

24. Richard A. Knox. "Researcher Disputes Location of Alzheimer's Gene Defect." *Boston Globe,* July 25, 1990, p. 3.

4. A BRIEF LOOK AT GENETICS

1. J. B. S. Haldane. *New Paths in Genetics.* New York: Harper and Brothers, 1942, p. 11.

2. Gregor Mendel. *Experiments in Plant Hybridisation.* [1865] Translation prepared by the Royal Historical Society of London, with notes by W. Bateson. Cambridge: Harvard University Press, 1950.

3. Thomas Hunt Morgan. "What Are 'Factors' in Mendelian Explanations?" *American Breeders Association,* vol. 5, 1909, p. 365.

4. Thomas Hunt Morgan, A. H. Sturtevant, H. J. Muller, and C. B. Bridges.

The Mechanism of Mendelian Heredity. New York: Henry Holt and Co., 1915, p. 210.

5. Thomas Hunt Morgan. *The Theory of the Gene.* New Haven: Yale University Press, 1926, p. 1.

6. Frederick Engels. *Herr Eugen Dühring's Revolution in Science (Anti-Dühring).* [1878] New York: International Publishers, 1939, p. 91.

7. Mani Mahadevan et al. "Myotonic Dystrophy Mutation: An Unstable CTG Repeat in the 3' Untranslated Region of the Gene." *Science,* vol. 255, 1992, pp. 1253–1255; Y.-H. Fu et al. "An Unstable Triplet Repeat in a Gene Related to Myotonic Muscular Dystrophy." *Science,* vol. 255, 1992, pp. 1256–1258.

8. Pamela Zurer. "Panel Plots Strategy for Human Genome Studies." *Chemical and Engineering News,* January 9, 1989, p. 5; Stephen S. Hall. "James Watson and the Search for Biology's 'Holy Grail.'" *Smithsonian Magazine,* February 1990, pp. 41–49.

9. F. Gianelli et al. "Haemophilia B: Data Base of Point Mutations and Short Additions and Deletions." *Nucleic Acid Research,* vol. 18, 1990, pp. 4053–4059.

10. Janice A. Egeland et al. "Bipolar Affective Disorders Linked to DNA Markers on Chromosome 11." *Nature,* vol. 325, 1987, pp. 783–787; John R. Kelsoe et al. "Re-evaluation of the Linkage Relationship Between Chromosome 11's Loci and the Gene for Bipolar Affective Disorder in the Old Order Amish." *Nature,* vol. 342, 1989, pp. 238–243.

5. Genes in Context

1. Leo Tolstoy. *The Death of Ivan Ilyich.* [1886] New York: Bantam Books, 1981.

2. Rainer Maria Rilke. *The Notebooks of Malte Laurids Brigge.* [1910] New York: Random House, 1983.

3. Sylvia Noble Tesh. *Hidden Arguments: Political Ideology and Disease Prevention Policy.* New Brunswick, New Jersey: Rutgers University Press, 1988.

4. Thomas McKeown. *The Modern Rise of Population.* New York: Academic Press, 1976.

5. Tesh. *Hidden Arguments,* p. 35.

6. Dolores Kong. "Age-old Illnesses Making Comeback." *Boston Globe,* September 9, 1991, pp. 41, 46.

7. Tesh. *Hidden Arguments,* p. 39.

8. Daniel E. Koshland, Jr. "Sequences and Consequences of the Human Genome." *Science,* vol. 246, 1989, p. 189.

9. Alexander Reid. "Death Rates Differ for Black, White Infants." *Boston Globe,* September 7, 1990, p. 13; Dolores Kong. "Black Infant Mortality Soars: Race, Economics Drive Boston Rates." *Boston Sunday Globe,* September 9, 1990, p. 1.

10. James P. Grant and The United Nations Children's Fund. *The State of the World's Children, 1990.* Oxford: Oxford University Press, 1990, p. 4.

11. Linda Rudolph et al. "Environmental and Biological Monitoring for Lead Exposure in California Workplaces." *American Journal of Public Health,* vol. 80, 1990, pp. 921–925.

12. Larry Tye. "Lead Poisoning Risk Greater than Thought." *Boston Globe,* July 19, 1990, p. 3.

13. Deborah Mesce. "Health Goals Are Set for the Year 2000; Prevention Stressed." *Boston Globe,* September 7, 1990, p. 3.

14. Gideon Koren and Naomi Klein. "Bias Against Negative Studies in Newspaper Reports of Medical Research." *Journal of the American Medical Association,* vol. 266, 1991, pp. 1824–1826.

15. Arno G. Motulsky. "Societal Problems in Human and Medical Genetics." *Genome,* vol. 31, 1989, pp. 870–875.

16. Malcolm Ritter. "Gene is Cloned in Search for Schizophrenia Drug." *Boston Globe,* September 6, 1990, p. 9.

17. Barry Worth. "How Short Is too Short?" *New York Times Magazine,* June 16, 1991, pp. 13–17, 28–29, 47.

18. John Lantos, Mark Siegler, and Leona Cuttler. "Ethical Issues in Growth Hormone Therapy." *Journal of the American Medical Association,* vol. 261, pp. 1020–1024.

19. Daniel Rudman et al. "Effects of Human Growth Hormone in Men Over 60 Years Old." *New England Journal of Medicine,* vol. 323, 1990, pp. 1–6.

20. Mary Lee Vance. "Growth Hormone for the Elderly?" *New England Journal of Medicine,* vol. 323, 1990, pp. 52–54.

6. "Inherited Tendencies": Chronic Conditions

1. Michael Swift, Daphne Morrell, Ruby B. Massey, and Charles L. Chase. "Incidence of Cancer in Families Affected by Ataxia-Telangiectasia." *New England Journal of Medicine,* vol. 325, 1991, pp. 1831–1836.

2. Timo E. Strandberg et al. "Long-term Mortality After 5-Year Multifactorial Primary Prevention of Cardiovascular Diseases in Middle-aged Men." *Journal of the American Medical Association,* vol. 266, 1991, pp. 1225–1229.

3. Richard A. Knox. "Study Finds Heart Deaths Rose Despite Care for Risks." *Boston Globe,* September 4, 1991, pp. 1, 6.

4. Francis Galton. *Hereditary Genius.* London: Macmillan, 1869.

5. Philip J. Hilts. "Agency Rejects Study Linking Genes to Crime." *New York Times,* September 6, 1992, p. 1.

6. G. I. Bell et al. "The Molecular Genetics of Diabetes Mellitus." In *Molecular Approaches to Human Polygenic Diseases.* Ciba Foundation Symposium, Chichester, England and New York: John Wiley and Sons, 1987, pp. 167–183.

7. Susan P. Helmrich, David R. Ragland, Rita W. Leung, and Ralph S. Paffenbarger. "Physical Activity and Reduced Occurrence of Non-Insulin-Dependent Diabetes Mellitus." *New England Journal of Medicine,* vol. 325, 1991, pp. 147–152.

8. Dolores Kong, "Exercise Can Help Prevent Diabetes, Large Study Says." *Boston Globe,* July 18, 1991, p. 9.

9. Helmrich et al. "Physical Activity and Reduced Occurrence of Non-Insulin-Dependent Diabetes Mellitus."

10. John A. Todd et al. "Genetic Analysis of Autoimmune Type 1 Diabetes Mellitus in Mice." *Nature,* vol. 351, 1991, pp. 542–547.

11. Thomas B. Newman, Warren S. Browner, and Stephen B. Hulley. "The Case Against Childhood Cholesterol Screening." *Journal of the American Medical Association,* vol. 264, 1990, pp. 3039–3043.

12. Richard Lewontin. *Human Diversity.* New York: Scientific American Books, 1982, chapter 8.

13. Thomas W. Wilson and Clarence E. Grim. "Biohistory of Slavery and Blood Pressure Differences in Blacks Today: A Hypothesis." *Hypertension,* vol. 17, 1991, supplement I, pp. 122–128.

14. Fatimah Linda Collier Jackson. "An Evolutionary Perspective on Salt, Hypertension, and Human Genetic Variability." *Hypertension,* vol. 17, 1991, supplement I, pp. 129–132.

15. Kathy A. Fackelmann. "The African Gene?" *Science News,* vol. 140, 1991, pp. 254–255.

16. Demetrius Albanes and Myron Winnick. "Are Cell Number and Cell Proliferation Risk Factors for Cancer?" *Journal of the National Cancer Institute,* vol. 80, 1988, pp. 772–775.

17. John Cairns. *Cancer: Science and Society.* San Francisco: W. H. Freeman and Co., 1978, p. 53.

18. Samuel S. Epstein. *The Politics of Cancer.* San Francisco: Sierra Club Books, 1978, p. 23; Richard Doll and Richard Peto. "The Causes of Cancer: Quantitative Estimates of Avoidable Risks of Cancer in the United States Today." *Journal of the National Cancer Institute,* vol. 66, 1981, pp. 1191–1308 (p. 1205).

19. Robert A. Weinberg. "A Molecular Basis of Cancer." *Scientific American,* vol. 249, November 1983, pp. 126–142; Robert A. Weinberg. "Finding the Anti-Oncogene." *Scientific American,* vol. 259, September 1988, pp. 44–51.

20. Junya Toguchida et al. "Prevalence and Spectrum of Germline Mutations of the p53 Gene Among Patients with Sarcoma." *New England Journal of Medicine,* vol. 326, 1992, pp. 1301–1308; David Malkin et al. "Germline Mutations of the p53 Tumor-Suppressor Gene in Children and Young Adults with Second Malignant Neoplasms." *New England Journal of Medicine,* vol. 326, 1992, pp. 1309–1315.

21. Cairns. *Cancer: Science and Society,* p. 22.

22. Ibid., p. 46.

23. Charles A. LeMaistre. "Reflections on Disease Prevention." *Cancer,* vol. 62, 1988, pp. 1673–1675.

24. Sandra Blakeslee. "Faulty Math Heightens Fears of Breast Cancer." *New York Times,* March 15, 1992, section 4, p. 1.

25. Ibid.

26. Ibid.

27. Susan M. Love, with Karen Lindsey. *Dr. Susan Love's Breast Book.* Reading, Mass.: Addison-Wesley, 1990, 1991, chapters 11 and 12.

28. Blakeslee. "Faulty Math Heightens Fears of Breast Cancer."

29. Nancy Krieger. "Exposure, Susceptibility, and Breast Cancer Risk." *Breast Cancer Research and Treatment,* vol. 13, 1989, pp. 205–223.

30. Blakeslee. "Faulty Math Heightens Fears of Breast Cancer."

31. J. E. Devitt. "False Alarms of Breast Cancer." *Lancet,* vol. 2, November 25, 1989, pp. 1257–1258.

32. Jean Marx. "Efforts to Prevent Cancer Are on the Increase." *Science,* vol. 253, 1991, p. 613; Judy Foreman. "U.S. to Begin a Wide Test of Breast Cancer Drug." *Boston Globe,* April 29, 1992, pp. 1, 12.

33. Janet Raloff. "Tamoxifen Quandary." *Science News,* vol. 141, April 25, 1992, pp. 266–269.

34. Devitt. "False Alarms of Breast Cancer."

35. Marx. "Efforts to Prevent Cancer."

36. Ann Gibbons. "Does War on Cancer Equal War on Poverty?" *Science,* vol. 253, 1991, p. 260.

37. I. Bernard Weinstein. "The Origins of Human Cancer: Molecular Mechanisms of Carcinogenesis and Their Implications for Cancer Prevention and Treatment—Twenty-seventh G. H. A. Clowes Memorial Award Lecture." *Cancer Research,* vol. 48, 1988, pp. 4135–4143.

7. "INHERITED TENDENCIES": BEHAVIORS

1. Michel Foucault. *The History of Sexuality, Volume 1: An Introduction.* New York: Vintage Books, 1980, p. 43.

2. Jeffrey Weeks. *Coming Out: Homosexual Politics in Britain from the Nineteenth Century to the Present.* London: Quartet Books, 1977, p. 62.

3. David Gelman with Donna Foote, Todd Barrett, and Mary Talbot. "Born or Bred?" *Newsweek,* February 24, 1992, pp. 46–53.

4. Ibid.

5. Anne Fausto-Sterling. *Myths of Gender: Biological Theories about Women and Men,* 2d edition. New York: Basic Books, 1992, chapter 8; B. Bower. "Gene Influence Tied to Sexual Orientation." *Science News,* vol. 141, January 4, 1992, p. 6.

6. Simon LeVay. "A Difference in Hypothalamic Structure Between Heterosexual and Homosexual Men." *Science,* vol. 253, 1991, pp. 1034–1037.

7. Fausto-Sterling. *Myths of Gender,* 2d edition, p. 252.

8. J. Michael Bailey and Richard C. Pillard. "A Genetic Study of Male Sexual Orientation." *Archives of General Psychiatry,* vol. 48, 1991, pp. 1089–1096.

9. Bower, "Gene Influence."

10. Gelman. "Born or Bred?" p. 49.

11. Constance Holden. "Twin Study Links Genes to Homosexuality." *Science,* vol. 255, 1992, p. 33.

12. B. Bower. "Genetic Clues to Female Homosexuality." *Science News,* August 22, 1992, p. 117.

13. Laura S. Allen and Roger A. Gorski. "Sexual Orientation and the Size of the Anterior Commissure in the Human Brain." *Proceedings of the National Academy of Sciences,* vol. 89, 1992, pp. 7199–7202.

14. Dean Hamer. *Biological Determinants of Human Sexuality.* Leaflet to solicit participants for an NIH study.

15. Lester Grinspoon and James B. Bakalar. "The Nature and Causes of Alcoholism." *The Harvard Medical School Mental Health Review,* no. 2, 1990, pp. 1–6.

16. Charles A. LeMaistre. "Reflections on Disease Prevention." *Cancer,* vol. 62, 1988, pp. 1673–1675; William Pollin. "The Role of the Addictive Process as a Key Step in Causation of All Tobacco-Related Diseases." *Journal of the American Medical Association,* vol. 252, 1984, p. 2874.

17. Herbert Fingarette. *Heavy Drinking: The Myth of Alcoholism as a Disease.* Berkeley: University of California Press, 1988.

18. Ernest Harburg et al. "Familial Transmission of Alcohol Use: II. Imitation of and Aversion to Parent Drinking (1960) by Adult Offspring (1977): Tecumseh, Michigan." *Journal of Studies on Alcohol,* vol. 51, 1990, pp. 245–256.

19. Christina Robb. "Alcoholism and Heredity: Study Finds Children of Heavy Drinkers Tend Toward Moderation." *Boston Globe,* May 28, 1991, p. 3.

20. Stanton Peele and Archie Brodsky, with Mary Arnold. *The Truth About Addiction and Recovery.* New York: Simon and Schuster (Fireside), 1992, p. 69.

21. Robb. "Alcoholism and Heredity."

22. Theodore Reich. "Biologic-Marker Studies in Alcoholism." *New England Journal of Medicine,* vol. 318, 1988, pp. 180–182.

23. Stanton Peele. "Second Thoughts About a Gene for Alcoholism." *Atlantic Monthly,* August 1990, pp. 52–58.

24. Donald W. Goodwin et al. "Alcohol Problems in Adoptees Raised Apart from Alcoholic Biological Parents." *Archives of General Psychiatry,* vol. 28, 1973, pp. 238–243.

25. Fingarette. *Heavy Drinking,* p. 52.

26. R. C. Lewontin, Steven Rose, and Leon J. Kamin. *Not In Our Genes: Biology, Ideology and Human Nature.* New York: Pantheon, 1984.

27. Donald W. Goodwin et al. "Alcoholism and Depression in Adopted-Out Daughters of Alcoholics." *Archives of General Psychiatry,* vol. 34, 1977, pp. 751–755.

28. Matt McGue, Roy W. Pickens, and Dace S. Svikis. "Sex and Age Effects on the Inheritance of Alcohol Problems: A Twin Study." *Journal of Abnormal Psychology,* vol. 101, 1992, pp. 3–17.

29. Bruce Bower. "Gene in the Bottle." *Science News,* vol. 140, 1991, pp. 190–191.

30. David E. Comings et al. "The Dopamine D_2 Receptor Locus as a Modifying Gene in Neuropsychiatric Disorders." *Journal of the American Medical Association,* vol. 266, 1991, pp. 1793–1800.

31. Joel Gelernter et al. "No Association Between an Allele at the D_2 Dopa-

mine Receptor Gene (DRD₂) and Alcoholism." *Journal of the American Medical Association,* vol. 266, 1991, pp. 1801–1807.

32. C. Robert Cloninger. "D₂ Dopamine Receptor Gene is Associated but Not Linked With Alcoholism." *Journal of the American Medical Association,* vol. 266, 1991, pp. 1833–1834.

33. Robert Plomin. "The Role of Inheritance in Behavior." *Science,* vol. 248, 1990, pp. 183–188.

34. Ibid.

35. Robert Plomin and Denise Daniels. "Why Are Children in the Same Family so Different from One Another?" *Behavioral and Brain Sciences,* vol. 10, 1987, pp. 1–60.

36. E. O'Callaghan et al. "Schizophrenia After Prenatal Exposure to 1957 A2 Influenza Epidemic." *Lancet,* vol. 337, 1991, pp. 1248–1249.

37. Richard L. Suddath et al. "Anatomical Abnormalities in the Brains of Monozygotic Twins Discordant for Schizophrenia." *New England Journal of Medicine,* vol. 322, 1990, pp. 789–794.

38. Fox Butterfield. "Studies Find a Family Link to Criminality." *New York Times,* January 31, 1992, p. A1.

39. Ruth Hubbard. *The Politics of Women's Biology.* New Brunswick, New Jersey: Rutgers University Press, 1990, especially chapters 9 and 11.

40. Patricia A. Jacobs et al. "Aggressive Behaviour, Mental Sub-normality and the *XYY* Male." *Nature,* vol. 208, 1965, pp. 1351–1352.

41. Jon Beckwith and Jonathan King. "The XYY Syndrome: A Dangerous Myth." *New Scientist,* vol. 64, 1974, pp. 474–476.

8. MANIPULATING OUR GENES

1. Steven A. Rosenberg et al. "Gene Transfer into Humans—Immuno-therapy of Patients with Advanced Melanoma, Using Tumor-Infiltrating Lymphocytes Modified by Retroviral Gene Transduction." *New England Journal of Medicine,* vol. 323, 1990, pp. 570–578; Denis Cournoyer and C. Thomas Caskey. "Gene Transfer into Humans: A First Step." *New England Journal of Medicine,* vol. 323, 1990, pp. 601–603.

2. Richard Knox. "4-Year-Old Gets Historic Gene Implant." *Boston Globe,* September 15, 1990, p. 1; Barbara J. Culliton. "ADA Gene Therapy Enters the Competition." *Science,* vol. 249, 1990, p. 975.

3. Deborah Erickson. "Genes to Order." *Scientific American,* vol. 266, June 1992, pp. 112–114.

4. Melissa A. Rosenfeld et al. "In Vivo Transfer of the Human Cystic Fibrosis Transmembrane Conductance Regulator Gene to the Airway Epithelium." *Cell,* vol. 68, 1992, pp. 143–155.

5. Jean Marx. "Gene Therapy for CF Advances." *Science,* vol. 255, 1992, p. 289.

6. Natalie Angier. "With Direct Injection, Gene Therapy Takes a Step into a New Age." *New York Times,* April 14, 1992, p. C3.

7. Edward M. Berger and Bernard M. Gert. "Genetic Disorders and the Ethical Status of Germ-Line Gene Therapy." *Journal of Medicine and Philosophy,* vol. 16, 1991, pp. 667–683.

8. Daniel E. Koshland, Jr. "The Future of Biological Research: What Is Possible and What Is Ethical?" *MBL Science,* vol. 3, 1988–1989, pp. 10–15.

9. Genes for Sale

1. Philip H. Abelson. "Biotechnology in a Global Economy." *Science,* vol. 255, 1992, p. 381.

2. Ann M. Thayer. "Biotech Companies in Good Shape Despite Large Losses in 1991." *Chemical and Engineering News,* March 30, 1992, pp. 9–11.

3. Charles Schwartz. "Corporate Connections of Notable Scientists." *Science for the People,* May 1975, pp. 30–31.

4. Samuel S. Epstein. *The Politics of Cancer.* San Francisco: Sierra Club Books, 1978, pp. 438–439.

5. Morton Mintz. *By Prescription Only.* Boston: Beacon Press, 1967, chapter 15; Ralph W. Moss. *The Cancer Syndrome.* New York: Grove Press, 1982, p. 300.

6. Christopher Anderson. "Conflict Concerns Disrupt Panels, Cloud Testimony." *Nature,* vol. 355, 1992, pp. 753–754.

7. Nachama L. Wilker. "Combatting Biotech's Corporate Virus: Conflict of Interest." *Genewatch,* vol. 7, November 1991, pp. 6–7.

8. Peter Gosselin. "Flawed Study Helps Doctors Profit on Drug." *Boston Globe,* October 19, 1988, pp. 1, 16–17.

9. Dolores Kong. "Charges against Two Dropped." *Boston Globe,* April 13, 1992, p. 50.

10. Sheldon Krimsky, James G. Ennis, and Robert Weissman. "Biotech Industry's Alliance with Scholars: Stronger, Deeper than Imagined." *Genewatch,* vol. 7, November 1991, pp. 1–2.

11. Ted Weiss. "Congress, Whistleblowers, and the Scientific Community: Can We Work Together?" *Genewatch,* vol. 7, November 1991, pp. 3–5.

12. Gale Scott. "Blood Test Predicts Some Breast Cancer." *Newsday,* April 28, 1992, p. 6.

13. Sheldon Krimsky. *Biotechnics and Society: The Rise of Industrial Genetics.* New York: Praeger, 1991.

14. Ronald Rosenberg and Jolie Solomon. "Seragen Reported to Weigh Cutting Share Offer Price." *Boston Globe,* March 28, 1992, p. 25.

15. Martin Kenney. *Biotechnology: The University-Industrial Complex.* New Haven: Yale University Press, 1986, p. 63.

16. Ibid., chapter 3.

17. Krimsky. *Biotechnics and Society,* p. 69.

18. Kenney. *Biotechnology,* p. 50.

19. Krimsky. *Biotechnics and Society,* p. 70.

20. Ronald Rosenberg. "Collaborative in First Public Offering with 1.5m Shares." *Boston Globe,* December 10, 1981, pp. 89–90.

21. No Author. "DNA Pioneer Quits Gene Map Project." *New York Times,* April 11, 1992, p. 12.

22. Wilker. "Combatting Biotech's Corporate Virus."

23. Elliot Diringer. "Reagan Biotech Adviser Faces Criminal Probe." *San Francisco Chronicle,* October 16, 1987, p. 1.

24. Mark Crawford. "Document Links NSF Official to Biotech Firm." *Science,* vol. 238, 1987, p. 742.

25. Weiss. "Congress, Whistleblowers, and the Scientific Community," p. 5.

26. Leslie Roberts. "Genome Patent Fight Erupts." *Science,* vol. 254, 1991, pp. 184–186.

27. Mark D. Adams et al. "Complementary DNA Sequencing: Expressed Sequence Tags and Human Genome Project." *Science,* vol. 252, 1991, pp. 1651–1656; Mark D. Adams et al. "Sequence Identification of 2,375 Human Brain Genes." *Nature,* vol. 355, 1992, pp. 632–634.

28. Roberts. "Genome Patent Fight Erupts," p. 184.

29. Bernadine Healy. "Special Report on Gene Patenting." *The New England Journal of Medicine,* vol. 327, 1992, pp. 664–668.

30. Roberts. "Genome Patent Fight Erupts," p. 184.

31. Christopher Anderson. "Controversial NIH Genome Researcher Leaves for New $70-million Institute." *Nature,* vol. 358, 1992, p. 95.

32. Healy. "Special Report on Gene Patenting," p. 667.

33. Christopher Anderson. "Gene Wars Escalate as US Official Battles NIH over Pursuit of Patent." *Nature,* vol. 359, 1992, p. 467.

34. No Author. "Free Trade in Human Sequence Data?" *Nature,* vol. 354, 1991, pp. 171–172; Richard Saltus. "Gene Patents: Weighing Protection vs. Secrecy." *Boston Globe,* December 2, 1991, pp. 25, 29.

10. Genetic Discrimination

1. Dorothy Nelkin and Laurence Tancredi. *Dangerous Diagnostics: The Social Power of Biological Information.* New York: Basic Books, 1989, especially chapter 6.

2. Frank R. Vellutino. "Dyslexia." *Scientific American,* vol. 256, March 1987, pp. 34–41.

3. Dorothy Nelkin and Laurence Tancredi. "Classify and Control: Genetic Information in the Schools." *American Journal of Law and Medicine,* vol. 17, 1991, pp. 51–73 (p. 67).

4. Ibid., p. 73.

5. Nelkin and Tancredi. *Dangerous Diagnostics,* p. 80.

6. Patrick Kinnersly. *The Hazards of Work: How to Fight Them.* London: Pluto Press, 1973, chapter 7.

7. Ruth Hubbard and Mary Sue Henifin. "Genetic Screening of Prospective Parents and of Workers: Some Scientific and Social Issues." In James M. Humber and Robert T. Almeder, eds. *Biomedical Ethics Reviews. 1984.* Clifton, New Jersey: Humana Press, 1984, pp. 73–120.

8. No Author. "More Genome Ethics." *Nature,* vol. 353, 1991, p. 2.

9. Council on Ethical and Judicial Affairs, American Medical Association. "Use of Genetic Testing by Employers." *Journal of the American Medical Association,* vol. 266, 1991, pp. 1827–1830.

10. Gilbert S. Omenn. "Predictive Identification of Hypersusceptible Individuals." *Journal of Occupational Medicine,* vol. 24, 1982, pp. 369–374.

11. U.S. Congress, Office of Technology Assessment. *Genetic Monitoring and Screening in the Workplace,* OTA-BA-455. Washington, D.C.: U.S. Government Printing Office, October 1990, p. 13.

12. Ibid., p. 166.

13. Larry Gostin. "Genetic Discrimination: The Use of Genetically Based Diagnostic and Prognostic Tests by Employers and Insurers." *American Journal of Law and Medicine,* vol. 17, 1991, pp. 109–144 (pp. 116–117).

14. Paul Billings. "Genetic Discrimination: An Ongoing Survey." *Genewatch,* vol. 6, nos. 4–5, no date, pp. 7, 15.

15. Hubbard and Henifin. "Genetic Screening."

16. U.S. Congress, Office of Technology Assessment. *Genetic Monitoring and Screening,* p. 14.

17. Ibid., p. 15.

18. Gostin. "Genetic Discrimination," p. 123.

19. Ibid., p. 132.

20. Ibid., pp. 142–143.

21. Ibid., p. 143.

22. Dorothy C. Wertz and John C. Fletcher. "An International Survey of Attitudes of Medical Geneticists Toward Mass Screening and Access to Results." *Public Health Reports,* vol. 104, 1989, pp. 35–44.

23. Bernard Shaw. *Selected Plays, with Prefaces.* New York: Dodd, Mead and Company 1948, vol. 1, p. 1.

24. Neil A. Holtzman. *Proceed With Caution.* Baltimore: Johns Hopkins University Press, 1989, p. 195.

25. Billings. "Genetic Discrimination."

26. Ibid., p. 15.

27. Paul Billings, Mel Kohn, Marguerite de Cuevas, Jon Beckwith, Joseph S. Alper, and Marvin R. Natowicz. "Discrimination as a Consequence of Genetic Testing." *American Journal of Human Genetics,* vol. 50, 1992, pp. 476–482.

28. Gostin. "Genetic Discrimination," p. 119.

29. See also, Jamie Stephenson. "A Case of Discrimination." *Genewatch,* vol. 7, February 1992, p. 9.

30. Gostin. "Genetic Discrimination," p. 135.

31. Wisconsin Senate Bill 483, Section 1121q.631.89, 1991.

11. DNA-BASED IDENTIFICATION SYSTEMS, PRIVACY, AND CIVIL LIBERTIES

1. Roger Lewin. "DNA Typing on the Witness Stand." *Science,* vol. 244, 1989, pp. 1033–1035.

2. Peter J. Neufeld and Neville Colman. "When Science Takes the Witness Stand." *Scientific American,* vol. 262, May 1990, pp. 46–53.

3. Eric S. Lander. "DNA Fingerprinting on Trial." *Nature,* vol. 339, 1989, pp. 501–505.

4. R. C. Lewontin and Daniel L. Hartl. "Population Genetics in Forensic DNA Typing." *Science,* vol. 254, 1991, pp. 1745–1750.

5. Janet C. Hoeffel. "The Dark Side of DNA Profiling: Unreliable Scientific Evidence Meets the Criminal Defendant." *Stanford Law Review,* vol. 42, 1990, pp. 465–538 (p. 493).

6. Lewontin and Hartl. "Population Genetics in Forensic DNA Typing."

7. Lander. "DNA Fingerprinting on Trial."

8. Ranajit Chakraborty and Kenneth K. Kidd. "The Utility of DNA Typing in Forensic Work." *Science,* vol. 254, 1991, pp. 1735–1739; Leslie Roberts. "Fight Erupts over DNA Fingerprinting." *Science,* vol. 254, 1991, pp. 1721–1723.

9. Christopher Anderson. "Conflict Concerns Disrupt Panels, Cloud Testimony." *Nature,* vol. 355, 1992, pp. 753–754.

10. Christopher Anderson. "DNA Fingerprinting Discord." *Nature,* vol. 354, 1991, p. 500.

11. Various Authors. "Letters: Forensic DNA Typing." *Science,* vol. 255, 1992, pp. 1050–1055.

12. National Research Council. *DNA Technology in Forensic Science.* Washington, D.C.: National Academy Press, 1992.

13. Gina Kolata. "U.S. Panel Seeking Restriction on Use of DNA in Courts." *New York Times,* April 14, 1992, pp. 1, C7.

14. Associated Press. "Genetic Data Reliable in Court, Panel Says." *Boston Globe,* April 15, 1992, p. 5.

15. Anderson. "DNA Fingerprinting Discord."

16. Anderson. "Conflict Concerns Disrupt Panels."

17. David P. Hamilton. "Letting the 'Cops' Make the Rules for DNA Fingerprints." *Science,* vol. 252, 1991, p. 1603.

18. Neufeld and Colman. "When Science Takes the Witness Stand."

19. Earl Ubell. "Whodunit? Quick, Check The Genes!" *Parade Magazine,* March 31, 1991, pp. 12–13.

20. 102d Congress, 1st session. H.R. 2045, April 24, 1991, p. 1.

21. Warren E. Leary. "Genetic Record to Be Kept on Members of Military." *New York Times,* January 12, 1992, p. 12.

22. U.S. Office of Personnel Management, Standard Form 85, Revised December 1987.

23. Alexander Dorozynski. "Privacy Rules Blindside French Glaucoma Effort." *Science,* vol. 252, 1991, pp. 369–370.

24. Philip L. Bereano. "DNA Identification Systems: Social Policy and Civil Liberties Concerns." Testimony before the Subcommittee on Civil and Constitutional Rights, Committee of the Judiciary, U.S. House of Representatives, March 22, 1989.

25. Joseph Wambaugh. *The Blooding.* New York: Bantam Books, 1989.

181

12. IN CONCLUSION . . .

1. "The Human Genome Project." A video released by the Public Affairs Department of the National Center for Human Genome Research of the National Institutes of Health, Bethesda, Maryland.
2. *General Electric Company* v. *Gilbert. Supreme Court Reporter,* vol. 97, 1976, pp. 401–421.
3. Hannah Arendt. *Eichmann in Jerusalem: A Report on the Banality of Evil.* New York: Penguin Books, 1977, p. 279.

APPENDIX

1. Lynn Margulis and Dorion Sagan. *Microcosmos: Four Billion Years of Microbial Evolution.* New York: Simon and Schuster, 1986, p. 128.
2. Lawrence I. Grossman. "Invited Editorial: Mitochondrial DNA in Sickness and Health." *American Journal of Human Genetics,* vol. 46, 1990, pp. 415–417.
3. I. J. Holt et al. "A New Mitochondrial Disease Associated with Mitochondrial DNA Heteroplasmy." *American Journal of Human Genetics,* vol. 46, 1990, pp. 428–433.
4. Grossman. "Invited Editorial."
5. Teri Randall. "Mitochondrial DNA: A New Frontier in Acquired and Inborn Gene Defects." *Journal of the American Medical Association,* vol. 266, 1991, pp. 1739–1740.
6. Simson L. Garfinkel. "Genetic Trails Lead to Argentina's Missing Children." *Boston Globe,* June 12, 1989, pp. 25, 27.
7. Christián Orego and Mary-Claire King. "Determination of Familial Relationships." In *PCR Protocols: A Guide to Methods and Applications.* New York: Academic Press, 1990, pp. 416–426.
8. Rebecca L. Cann, Mark Stoneking, and Allan C. Wilson. "Mitochondrial DNA and Human Evolution." *Nature,* vol. 325, 1987, pp. 31–36; Rebecca L. Cann. "In Search of Eve." *The Sciences,* September/October 1987, pp. 30–37.
9. Henry Gee. "Statistical Cloud over African Eden." *Nature,* vol. 355, 1992, p. 583; Alan G. Thorne and Milford H. Wolpoff. "The Multiregional Evolution of Humans." *Scientific American,* vol. 266, April 1992, pp. 76–83.

GLOSSARY

Acromegaly A syndrome characterized by exaggerated growth of facial features, hands, and feet, which results from an oversecretion of pituitary growth hormone.

Adenine One of the bases in DNA.

Adenosine deaminase (ADA) An enzyme involved in essential chemical transformations of the bases that are part of the structure of DNA. An inherited deficiency in ADA can result in defects of the immune system that lead to an increased susceptibility to various infections.

AIDS (Acquired Immunodeficiency Syndrome) A group of bacterial and parasitic infections that can take hold when a person's resistance has been weakened through infection by the human immunodeficiency virus (HIV).

Albumin Any one of a family of relatively small, water-soluble proteins that are found in egg white, blood, and other tissues.

Alcoholics Anonymous (AA) A self-help organization that enables alcoholics to meet together and support each other in their efforts to stop drinking and make other important changes in their lives.

Allele One of several possible forms of a gene, found at the same location on a chromosome, which can give rise to noticeable hereditary differences.

Amino acid Any one of a type of small molecules that constitute the building blocks of proteins.

Amniocentesis A surgical procedure in which a syringe is used to draw a sample of the fluid surrounding the growing fetus out of a woman's uterus, so that physicians can examine this fluid for cells that the fetus has shed into this fluid.

Anti-oncogene A gene that can suppress uncontrolled cell division and therefore prevent cancerous growth.

Arthritis A chronic, painful, and often progressive inflammation of the joints that is thought to originate in an allergy-like overresponsiveness of a person's immune system.

Atherosclerosis A thickening and hardening of the walls of the arteries due to the deposition of fatty substances.

Autosomes Those chromosomes in the cell nucleus which are not the so-called sex chromosomes, X and Y.

Base see *nucleotide*

Breeding true Producing offspring that conform to the ancestral type or stock.

Cancer A group of cells that have escaped the normal regulation that governs the rates of cell division in various tissues, which multiply at an uncontrolled pace and often can spread to tissues other than the one in which those cells originated.

Carcinogen A cancer-promoting agent, such as some chemicals, radiation, tobacco, or asbestos.

Carrier Someone who carries a recessive allele without exhibiting the corresponding trait.

Cell A unit of structure and function in different organs and organisms that is anatomically distinct by virtue of the fact that it is enclosed in a membrane.

Cell division The process by which a cell gives rise to two or more copies of itself.

Charcot-Marie-Tooth Disease A hereditary condition of the nervous system with variable manifestations, in which there may be weakness or some atrophy of the muscles of the extremities, especially the legs.

Cholesterol A fatty substance that produces the thickening of the arterial walls seen in atherosclerosis.

Chromosomes Microscopic structures, found in the cell nucleus and in mitochondria, which are composed of DNA and proteins and are duplicated every time a cell divides. The DNA in the chromosomes carries the genes.

Codon A sequence of three nucleotide bases that "codes" for a specific amino acid during the process in which the base sequence of a gene gets translated into a sequence of amino acids in a protein.

Congestive heart failure An incapacity of the heart muscle in which it fails to contract sufficiently to empty the heart of blood. This leads to progressively greater stress and weakening of the muscle until it finally ceases to contract.

Crossing-over A process that can occur during cell division, in which the two members of a pair of chromosomes that have been inherited from the two parents exchange corresponding parts of their DNA.

Cystic fibrosis An inherited condition in which a thick mucus tends to build up in the lungs and other tissues. This increases susceptibility to infections and other serious health problems.

Cytoplasm The part of the cell that lies between the membrane and the nucleus. It contains a variety of subcellular structures that participate in the cell's functions.

Cytosine One of the bases in DNA.

Deoxyribonucleic acid (DNA) The molecule in the chromosomes that specifies the composition of proteins. It is made up of a repeating sequence of one of four possible bases, a phosphate molecule, and a sugar (called deoxyribose), wound up into a helix.

Diabetes A disturbance of carbohydrate metabolism that results in the excretion of sugar in the urine and that can lead to various disorders, especially of the circulatory and nervous system.

DNA see *deoxyribonucleic acid*

DNA-fingerprinting A set of techniques intended to identify individuals, based on the base sequence of their DNA.

Dominant trait A trait that is apparent even if only one parent contributes the allele associated with it.

Double helix The usual geometric configuration of DNA, consisting of two complementary strands, each made up of a long, repeating sequence of sugar and phosphate molecules, running side by side in a helical formation, and joined by the bases.

Down syndrome A form of mental retardation of variable extent that is usually not hereditary and is associated with the presence of an extra copy of a normal chromosome (chromosome 21) in the nucleus of a person's cells.

Ectrodactyly An inherited condition in which some of the bones in the hands and feet may be fused and which therefore limits the mobility of the fingers and toes.

Edema A swelling produced by the accumulation of water and salts in the spaces surrounding the cells and tissues.

Enzyme Any one of a large class of proteins occurring in organisms, which make it possible for chemical reactions to take place and occur sufficiently rapidly to meet the organism's needs.

Epithelia The thin layers of cells lining the surfaces of tissues, such as the skin, the respiratory passages in the lungs, or the walls of the intestines, where the tissues come in contact with the external or internal environment.

Eugenics   A social theory that holds out the promise of improving the genetic endowment of human beings by encouraging people with "more desirable" traits to reproduce and discouraging or preventing those with "less desirable" ones from doing so.

Exon   A base sequence of DNA that gets translated into the amino acid sequence of a protein.

Fragile X syndrome   A recessive condition that can involve mental retardation as well as some physical symptoms, in which a section of three bases on the X-chromosome is repeated over and over. The number of repeats varies in different individuals and seems to be correlated with the extent of the symptoms.

Galactosemia   An inherited disorder due to the absence of an enzyme involved in carbohydrate metabolism, which can produce stunting of growth and mental retardation in children.

Gamete   A reproductive cell, such as an egg or a sperm.

Gaucher disease   An uncommon inherited disorder of fat metabolism in which fatty substances accumulate in various tissues.

Gene   A functional unit of DNA that specifies the composition of a protein and can be passed on from an individual to his or her descendants.

Genetic counseling   Counseling provided by a trained genetic counselor or medical geneticist, in which people are informed about implications of their biological family's health history or of genetic tests for their own health or that of their children.

Genetic linkage   The property of some traits or genes to be inherited together, which is interpreted to mean that these genes are located near each other on the same chromosome.

Genetic monitoring   Genetic testing, usually associated with a suspected exposure to agents that can cause gene mutations, used to determine whether such mutations have taken place.

Genetic screening   Population-based testing for alleles or markers associated with traits that are assumed to be inherited, irrespective of whether people have reason to think they may in fact have inherited the gene mutation for which they are being screened.

Geneticization   The attribution of genetic significance to characteristics that may not have any.

Genome   All the genes in each cell of an organism.

Genotype   An organism's genetic make-up.

Germ-line gene therapy   The attempt to insert genes into the nucleus of sperm, eggs, or early embryos, by which these new genes will then become part of the genetic endowment of the individual who develops out of that sperm, egg, or embryo as well as of her or his biological descendants.

Germ plasm   That part of the cell that is transmitted from one generation to the next.

Glaucoma   A condition in which the fluid pressure inside the eye increases, which may result in impaired vision and eventual blindness.

Globin   The colorless protein component of the red blood pigment hemoglobin.

Glucose   A "simple" sugar, which is a constituent of ordinary table sugar (sucrose) and of starch.

Guanine   One of the bases in DNA.

Hardy-Weinberg law   A theorem of population biology that defines the statistical relationship between the number of individuals in a randomly breeding population who manifest a recessively inherited trait and the number of individuals who carry a single copy of the relevant allele and therefore do not exhibit the trait.

Health maintenance organization (HMO)   A medical organization that delivers health care on the basis of an insurance plan in which subscribers pay a flat fee in advance, which then entitles them to all needed medical services.

Heme   The pigment responsible for the red color of hemoglobin.

Hemoglobin   The protein that is the main constituent of red blood cells, containing alpha- and beta-globin, heme, and iron. It picks up oxygen in the lungs and transports this oxygen to the rest of the body.

Hemophilia   An inherited failure of blood to clot that is due to the absence of one of several proteins, called clotting factors.

Hemophilia B   One form of hemophilia.

Heterozygous   Individuals are said to be heterozygous for a specific gene if they have inherited different alleles of that gene from their two parents.

HIV (human immunodeficiency virus)   A virus that attacks the immune system and renders it progressively less able to ward off infections.

Hominids   Members of the primate family Hominidae, of which modern humans are the only surviving species.

Homo sapiens   (latin for "wise man") Modern human beings, the only species of the primate family Hominidae now living.

Homozygous   Individuals are said to be homozygous for a particular gene if they inherited identical alleles of that gene from both their parents.

Human growth hormone   A protein hormone, secreted by the pituitary gland, which stimulates growth of bones and muscles.

Huntington disease   An inherited condition involving disorientation and progressive mental deterioration which usually begins to exhibit its symptoms as people move into their forties and fifties.

Hypertension   A chronic elevation of blood pressure above the level considered to be normal.

Iatrogenic   Induced by a doctor's words or actions.

Insulin   The protein hormone produced by the pancreas that regulates the glucose level in the blood.

Insulin receptor   A protein on cell surfaces that mediates the entry of insulin into cells where it can exert its metabolic effects.

Intron   A base sequence of DNA within a gene that gets eliminated before the rest of the gene is translated into the amino acid sequence of a protein.

Keratin   An insoluble and rather rigid protein that is the main structural component of hair, feathers, claws, and fingernails.

Lactose intolerance   An inherited lack of one or more intestinal enzymes that produces difficulties in digesting carbohydrates and can lead to diarrhea and painful bloating of the abdomen.

Linkage maps   Maps of the genes on a chromosome that are constructed on the basis of experimental observations showing that some traits tend to be inherited together. The assumption made in constructing such maps is that the genes corresponding to these traits lie close to each other.

Lipid   A class of fatty substances.

Lipoprotein   The combination of fatty substances (lipids) with a member of a specific group of proteins.

Lymphocyte   A kind of white blood cell that participates in the organism's immune response following an infection.

Mammogram   An x-ray examination of the breast.

Marker   An identifiable piece of DNA that lies on a chromosome near the unidentified segment that specifies the protein relevant to a certain trait.

Marker DNA   The piece of DNA that includes both the marker and the unidentified segment that specifies the trait.

Mean   The average of a range of values.

Meiosis   The type of cell division that occurs in the formation of eggs or sperm, during which the pairs of chromosomes in the parent cell separate and only one of each pair goes into each daughter cell.

Melanoma   A cancer that originates in pigment cells found in some areas of the skin, mucous membranes, eyes, and central nervous system.

Mendelian trait   A characteristic whose pattern of inheritance can be described using Mendel's laws.

Messenger-RNA (m-RNA)   A kind of RNA molecule that conveys the message encoded in specific base sequences of DNA to the particles in the cytoplasm where that base sequence will be translated into the amino acid sequence of the corresponding protein.

Metastases   Invasions of cancer cells into tissues other than the tissue in which they originated.

Mitochondria   Organized structures in the cell's cytoplasm involved in the processes by which cells transform foodstuffs to generate energy. Mitochondria contain their own complement of DNA, which is distinct from the DNA in the chromosomes of the cell nucleus.

Mitochondrial DNA   The DNA in the mitochondrial chromosomes.

Mitosis   The usual type of cell division, in which a cell gives rise to two cells that are identical to the original cell and to each other.

Monitoring   see *genetic monitoring*

Mutagen   An agent, such as radiation or chemicals, that increases the incidence of genetic mutations.

Mutation   A permanent alternation in a gene or DNA molecule.

Normal distribution   A symmetrical, bell-shaped curve that falls off to zero on both sides of the mean.

Nuclear DNA   The chromosal DNA in the nucleus of a cell.

Nucleotide   A subunit of DNA or RNA, made up of a base, a phosphate, and a sugar molecule.

Nucleotide base   see *nucleotide*

Nucleus   The part of a cell that contains the chromosomes (except the mitochondrial chromosomes) and the bulk of the cell's DNA.

Oncogene   A gene that has undergone a mutation, making it induce cells to divide more often than they normally would, which can lead to the growth of a cancer.

Oncologist   A physician who specializes in treating cancers.

Pellagra   A condition characterized by a range of physical and mental symptoms, which arises from a deficiency of niacin, a member of the vitamin B complex.

Phenotype   An organism's outward appearance.

Phenylalanine   An amino acid present in many proteins (see PKU).

Pituitary dwarfism   Abnormally short stature due to insufficient secretion of growth hormone by the pituitary gland.

PKU (phenylketonuria)   An inherited condition in which a deficiency in a specific enzyme results in the accumulation of a toxic derivative of the amino acid phenylalanine. This can be avoided by eliminating foods containing phenylalanine from the diet. Untreated PKU can result in mental retardation and neurological problems.

Polar body   A particle that is extruded and lost during the cell divisions involved in the formation of egg cells.

Polygenic trait   A trait whose transmission is thought to involve more than one gene.

Polymerase chain reaction (PCR)   A laboratory procedure in which enzymes are used to copy a tiny amount of DNA over and over until the sample is sufficiently large for chemical analysis or experimentation.

Polymorphisms   Different forms of the same trait or organism.

Protein   A large molecule composed of amino acid molecules strung end to end. Proteins participate in all the biological functions of cells and organisms.

Proto-oncogene   A gene capable of undergoing a mutation that can transform it into an oncogene.

Recessive trait   A trait that cannot become noticeable unless both parents contribute the allele associated with it.

Reduction division   Synonymous with meiosis.

Reductionism   The philosophical belief that phenomena or organisms are best understood by breaking them up into smaller parts.

Restriction enzyme   A type of enzyme that cuts DNA wherever there is a specific sequence of about four to six bases.

Restriction fragments   Sequences of DNA bases that are produced when restriction enzymes have chopped DNA into pieces.

Reverse transcriptase   An enzyme that can synthesize DNA from templates offered by RNA.

RFLPs (restriction fragment length polymorphisms)   Variations in the lengths of restriction fragments brought about by changes in the base sequence of DNA, or mutations, that alter the way restriction enzymes cut the DNA molecule.

Ribosomes   The particles in the cell cytoplasm on which the base sequences brought there by messenger-RNA get translated into the amino acid sequences of proteins.

RNA (ribonucleic acid)   A kind of nucleic acid that differs from DNA in that it contains the nucleotide base uracil, rather than thymine, and the sugar ribose instead of deoxyribose.

Screening   see *genetic screening*

Sex chromosomes   The X and Y chromosomes. Women have two X chromosomes, while men have one X and one Y.

Sickle-cell anemia   A recessively inherited condition in which the blood protein hemoglobin is altered so that it tends to clog the capillary blood vessels. This can lead to blood loss, painful joints, and infections.

Sickle-cell trait   The carrier state for sickle-cell anemia. It increases the carrier's resistance to malaria, but appears to be associated with no other symptoms.

Somatic gene therapy The attempt to modify the way certain tissues of an organism function by inserting a gene into cells of those tissues.

Symbiotic A close relationship between two organisms that may, but need not, benefit one or both of them.

Tamoxifen A chemical that counteracts the biological action of the hormone estrogen. It can be useful in the treatment of types of breast cancer whose growth is promoted by estrogen, but has a range of deleterious effects.

Tay-Sachs disease A recessively inherited condition that involves progressive slowing in development of young children and leads to paralysis, mental retardation, blindness, and finally death by the third or fourth year.

Thymine One of the bases in DNA.

Trait A specifiable characteristic of an organism, which may or may not be heritable.

Tumor suppressor gene Synonymous with anti-oncogene.

Tyrosine An amino acid that occurs in many proteins.

Vector A virus or other piece of DNA into which a gene can be incorporated and which can then be used to introduce that gene into a cell.

Virus A submicroscopic infectious agent, consisting of a core of DNA or RNA surrounded by protein, which is replicated inside living cells.

VNTRs (variable number of tandem repeats) Short base sequences on the chromosomes that are repeated over and over for a variable number of times. The reason such repeats occur is not known, but since the number of the repeats tends to differ among individuals, it can be used to identify specific individuals.

X chromosome One of the two so-called sex chromosomes. Ordinarily, females have two X chromosomes and males have one X and one Y.

Y chromosome One of the two so-called sex chromosomes, found ordinarily only in males.

Zygote A fertilized egg.

BOOKS AND OTHER RESOURCES

This list of resources is by no means comprehensive. I have limited myself to recently published books that I have found useful and interesting, and to science magazines and newsletters that regularly carry articles written for general readers. The list of organizations is intended to provide access to some groups that are trying to prevent or ameliorate inequities growing out of the present preoccupation with inherited "defects" and explore legal or legislative remedies. These lists are arranged alphabetically, and not in order of importance.

BOOKS

Arendt, Hannah. *Eichmann in Jerusalem: A Report on the Banality of Evil* (New York: Penguin, 1977). Based on her observations while attending the trial of Adolf Eichmann, Arendt reflects on the way unspeakable cruelties can be legally sanctioned and become accepted and ordinary.

Beinfield, Harriet and Efrem Korngold. *Between Heaven and Earth: A Guide to Chinese Medicine* (New York: Ballantine Books, 1991). Part theoretical exposition, part self-help manual, this book introduces readers to contemporary practices based on the ancient traditions of Chinese medicine, and hence to a very different view of illness, disease prevention, therapy, and cure from the one most of us grow up with in the Americas and Europe.

Cohen, Sherrill and Nadine Taub, eds. *Reproductive Laws for the 1990s* (Clifton, New Jersey: Humana Press, 1989). A collection of essays about the effects of the new developments in genetics and reproductive technologies on women's experience of childbearing, and about legal and legislative changes needed to reduce inequities and a narrowing of choice. Two introductory articles are especially important. One, by Laurie Nsiah-Jefferson, is about the impact of reproductive laws on low-income women and women of color, and the other, by Adrienne Asch, is about the complex implications of the new reproductive technologies for women with disabilities and for women who receive a prediction of disability for the fetus they are carrying.

Draper, Elaine. *Risky Business: Genetic Testing and Exclusionary Practices in the Hazardous Workplace* (New York: Cambridge University Press, 1991). A wide-ranging exploration of the use of genetic tests to limit workers' options or refuse them employment.

Duster, Troy. *Backdoor to Eugenics* (New York: Routledge, 1990). A warning about how current and potential uses of the ever-expanding list of tests for so-called genetic conditions will open a "backdoor" to eugenic practices even if the front door to eugenics is kept shut.

Gould, Stephen Jay. *The Mismeasure of Man* (New York: W. W. Norton, 1981). An illuminating discussion of the ways scientists in the nineteenth and twentieth centuries have produced and used numerical and other scientific data to bolster the existing prejudices of their societies. Gould's case histories are especially interesting because in most cases the scientists are not trying to mislead. Rather, their critical faculties are limited by their unquestioning acceptance of conventional assumptions.

Groce, Nora Ellen. *Everyone Here Spoke Sign Language: Hereditary Deafness on Martha's Vineyard* (Cambridge: Harvard University Press, 1985). An exploration of the way inhabitants of Martha's Vineyard, where a form of hereditary deafness was prevalent until some time in the first half of this century, so completely accepted this situation that the deafness ceased to be a "disability."

Holtzman, Neil A. *Proceed with Caution: Predicting Genetic Risks in the Recombinant DNA Era* (Baltimore: Johns Hopkins University Press, 1989). A discussion of the scientific basis of genetic diagnosis and the uncertainties and dangers of genetic predictions, by a professor of pediatrics and public health.

Hubbard, Ruth. *The Politics of Women's Biology* (New Brunswick, New Jersey: Rutgers University Press, 1990). A discussion of the ways societal assumptions and scientific explorations of women's biology bolster each other to clothe old prejudices in new guises.

Kevles, Daniel J. *In the Name of Eugenics: Genetics and the Uses of Human Heredity* (New York: Alfred A. Knopf, 1985). A readable history of eugenic practices in Great Britain and the United States from the mid-nineteenth century until the early 1980s.

Kevles, Daniel J. and Leroy Hood, eds. *The Code of Codes: Scientific and Social Issues in the Human Genome Project* (Cambridge: Harvard University Press, 1992). A collection of articles by social and natural scientists whose views about the Human Genome Project range from devout enthusiasm to critical questioning of its potential effects and effectiveness.

Krimsky, Sheldon. *Biotechnics and Society: The Rise of Industrial Genetics* (New York: Praeger, 1991). An overview of the first decade of biotechnology, the difficulties of adequate risk prediction, and the conflicts of interest that inevitably arise when supposedly disinterested biological and medical researchers become partners in for-profit industrial enterprises.

Lane, Harlan. *The Mask of Benevolence: Disabling the Deaf Community* (New York: Knopf, 1992). A strong indictment of the medicalization of deafness, especially new technological devices that make deaf people "hear." The author argues that deaf people constitute a linguistic minority and depriving deaf children of the opportunity to learn sign language and to have ongoing contact with deaf, signing adults seriously limits their social and intellectual development.

Lerner, Richard M. *Final Solutions: Biology, Prejudice, and Genocide* (University Park, Penn.: Pennsylvania State University Press, 1992). His analysis of the dreadful consequences of biological determinism leads the author to insist that human behavior be viewed as a dynamic interaction between heredity and environment, biology and culture.

Lewontin, Richard. *Human Diversity* (New York: Scientific American Books, 1982). A discussion of the genetics of human populations and the variations within and between them, as well as the evolutionary significance of human diversity.

Lewontin, R.C., Steven Rose, and Leon J. Kamin. *Not in Our Genes: Biology, Ideology, and Human Nature* (New York: Pantheon, 1984). An analysis of the ways genetics has been used to legitimate inequalities between long-standing inhabitants and immigrants, different racial groups, women and men, and other groups facing each other across socially erected gulfs of difference.

Lifton, Robert J. *The Nazi Doctors* (New York: Basic Books, 1986). An American psychiatrist's documentation of the extent to which German physicians accepted genocidal tasks as part of their medical mission, based on interviews with concentration camp doctors and their assistants, colleagues, and family members.

Miringoff, Marque-Luisa. *The Social Costs of Genetic Welfare* (New Brunswick, New Jersey: Rutgers University Press, 1991). An exploration of the contradictions between a social perspective that concentrates on improving people's lives, whether or not they have obvious disabilities, and a genetic focus that urges us to prevent people with disabilities from procreating or being born.

Müller-Hill, Benno. *Murderous Science* (Oxford and New York: Oxford University Press, 1988). A German geneticist's account of the involvement of well-known geneticists, anthropologists, and psychiatrists in the Nazi atrocities.

Nelkin, Dorothy and Laurence Tancredi. *Dangerous Diagnostics: The Social Power of Biological Information* (New York: Basic Books, 1989). An analysis of some consequences of biological testing and labeling of people in schools, workplaces, courts of law, and other significant contexts in our society.

Pope, Andrew M. and Alvin R. Tarlov. *Disability in America: Toward a National Agenda for Prevention* (Washington, D.C.: National Academy Press, 1991). A description of the conclusions reached by a committee of the National Academy of Sciences looking into disabilities arising from injury, malfunctions of development, chronic health conditions, and aging, which proposes measures to minimize the effects of such conditions on people's lives.

Proctor, Robert N. *Racial Hygiene: Medicine under the Nazis* (Cambridge: Harvard University Press, 1988). An analysis of the role of scientists and physicians in the development and implementation of the Nazi extermination programs, by an American historian of science.

Stone, Deborah A. *The Disabled State* (Philadelphia: Temple University Press, 1984). A provocative exploration of our culture's approach to disabilities and of the measures required to counteract societal prejudices.

Suzuki, David and Peter Knudtson. *Genethics: The Ethics of Engineering Life* (Toronto: Stoddart, 1988). A discussion of modern genetics and some of its social implications.

Tesh, Sylvia Noble. *Hidden Arguments: Political Ideology and Disease Prevention Policy* (New Brunswick, New Jersey: Rutgers University Press, 1988). An analysis of the political basis of most contemporary discussions about health, illness, and disease prevention.

U.S. Congress, Office of Technology Assessment. *Genetic Monitoring and Screening in the Workplace,* OTA-BA-455 (Washington, D.C.: U.S. Government Printing Office, October 1990). Results of a survey of the uses of genetic tests to monitor and screen employees, conducted by the Office of Technology Assessment. Unfortunately, the survey suffers from the low return rate of questionnaires sent out to employers.

U.S. Congress, Office of Technology Assessment. *Genetic Witness: Forensic Uses of DNA Tests*, OTA-BA-483 (Washington, D.C.: U.S. Government Printing Office, July 1990). An overly optimistic assessment of the benefits of DNA testing in the context of criminal proceedings, which underestimates both technical problems and likely encroachments on privacy and civil liberties.

U.S. Congress, Office of Technology Assessment. *Medical Testing and Health Insurance,* OTA-H-384 (Washington, D.C.: U.S.Government Printing Office, August 1988). A survey of the status and consequences of testing by health insurers and employers for actual and predicted diseases.

Yoxen, Edward. *The Gene Business: Who Should Control Biotechnology?* (New York: Harper and Row, 1983). An exploration of the consequences of turning molecular biology into an industry.

Zola, Irving Kenneth. *Missing Pieces: A Chronicle of Living With a Disability* (Philadelphia: Temple University Press, 1982). An autobiographical account by an American sociologist and disability rights activist of time spent as a participant observer in a community in the Netherlands in which everyone uses a wheelchair.

MAGAZINES

American Journal for Human Genetics. A monthly journal, published by the American Society for Human Genetics, that focuses on technical articles, but sometimes carries editorials and review articles of general interest.

Genetic Resource (150 Tremont Street, Boston, MA 02111). A magazine published twice a year by the Genetics Program of the Massachusetts Department of Public Health. With articles about new developments in genetics, prenatal diagnosis, biotechnology, and related areas.

Genewatch (19 Garden Street, Cambridge, MA 02138). The bimonthly newsletter of the Council for Responsible Genetics. *Genewatch* publishes editorials, articles, and book reviews about various aspects of genetics and biotechnology, intended for a general audience.

Hastings Center Report (255 Elm Road, Briarcliff Manor, NY 10510). A bimonthly newsletter with articles and commentaries about bioethics, which frequently deals with issues arising from genetic research and the genome initiative.

Human Genome News. (National Center for Human Genome Research, National Institutes of Health, Bethesda, MD 20892). A bimonthly newsletter that summarizes news and events relating to the Human Genome Project.

Lancet. A weekly medical magazine, published in London and Baltimore, that often carries articles and letters of general interest.

Nature. A weekly scientific magazine, published in London, that carries editorials and a science news section, along with specialized articles intended for a technical readership.

New England Journal of Medicine. A weekly medical journal, published by the Massachusetts Medical Society, that carries editorials and review articles of general interest, as well as specialized, technical articles.

New Scientist. A London-based weekly that publishes news articles and commentaries about scientific developments, written for a general audience.

Science. A journal published once a week in Washington, D.C., by the American Association for the Advancement of Science. Though most of *Science* consists of technical articles, each issue carries an editorial as well as news sections that summarize that week's events in nontechnical language.

Science News. A weekly magazine, intended for a general readership, published by Science Service in Washington, D.C.

Scientific American. A monthly magazine, published in New York and directed at a general readership, that often carries news and articles about genetics, genetic diagnosis, health, and public policy.

ORGANIZATIONS

American Civil Liberties Union (132 West 43d Street, New York, NY 10036). A national organization, with chapters in many states, which is becoming increasingly interested in the implications of the new genetic technologies for privacy, civil liberties, and the rights of defendants in criminal trials and of incarcerated people.

American Society of Law and Medicine (765 Commonwealth Avenue, Boston, MA 02215). An organization that holds meetings and publishes a journal, both of which often address legal and social issues arising from research in genetics.

Council for Responsible Genetics (19 Garden Street, Cambridge, MA 02138). The Council is a national organization of scientists, health professionals, trade unionists, women's health activists, and others who want to make sure that biotechnology is developed safely and in the public interest. The Council publishes a newsletter, *Genewatch*, and has available position papers on the Human Genome Initiative, genetic discrimination, germ-line gene modifications, and DNA-based identification systems.

Disability Rights Education and Defense Fund (1633 Q Street, Washington, D.C. 20009). A national organization of disability rights activists, interested in the implications of the new genetic technologies for people with disabilities and their families.

Genetic Screening Study Group (c/o Prof. Jon Beckwith, Department of Microbiology, Harvard Medical School, Boston, MA 02115). A Boston-based group of biomedical researchers exploring the social and scientific implications of the new genetic technologies, whose members publish articles and organize conferences intended for journalists and the public as well as for scientists.

National Center for Education in Maternal and Child Health (38th and R Streets, NW, Washington, D.C. 20057). A center that lists organizations that publish newsletters and other materials, intended for people who want to contact advocacy groups concerned with specific health conditions.

National Women's Health Network (1325 G Street, N.W., Washington, D.C. 20005). A national advocacy and lobbying organization, concerned with issues affecting the health of women. Publishes a bimonthly newsletter and periodic alerts and policy statements.

Project on Women and Disability (1 Ashburton Place, Room 1305, Boston, MA 02108). A division of the Massachusetts Office of Handicapped Affairs that tries to collect information and educate people with disabilities about issues of genetic discrimination.

INDEX